In-Memory Data Management Research

Series Editor

Prof. Dr. Dr. h.c. Hasso Plattner
Hasso Plattner Institute
Potsdam, Germany

For further volumes:
http://www.springer.com/series/11642

This book series presents selected research results in the context of In-Memory Data Management. The volumes in this series describe research results in in-memory database technology, logical and physical data management, software architectures, real-time analysis of enterprise data, innovative new business applications, and influenced business processes. In addition, programming models and software engineering techniques, tools, and benchmarks are elaborated on and discussed. All books are introduced by a member of the editorial board, who outlines the popular context and the social relevance of each work.

Globally, companies generate a steadily increasing amount of data, day after day. This data is obtained to optimize logistics, create knowledge, explore business relationships, and to improve management decisions. The trend towards acquiring more and more data, also known as "big data," requires fundamental support in data analysis. In-Memory Technology – the common theme in all volumes of this series – has become a de facto standard for fulfilling new requirements that are stated towards enterprise applications.

Hasso Plattner • Matthieu-P. Schapranow
Editors

High-Performance In-Memory Genome Data Analysis

How In-Memory Database Technology Accelerates Personalized Medicine

Springer

Editors
Hasso Plattner
Hasso Plattner Institute
Enterprise Platform and Integration Concepts
Potsdam, Germany

Matthieu-P. Schapranow
Hasso Plattner Institute
Enterprise Platform and Integration Concepts
Potsdam, Germany

ISBN 978-3-319-03034-0 ISBN 978-3-319-03035-7 (eBook)
DOI 10.1007/978-3-319-03035-7
Springer Cham Heidelberg New York Dordrecht London

Libarary of Congress Control Number: 2013954438

Printed on acid-free paper

Springer is part of Springer Science+Business Media (www.springer.com)

Quotes

“An increased utility of sequencing data should follow from the ability to process hundreds of gigabytes of raw sequence data informatically prior to subsequent downstream analysis. Plattner and Schapranow share concrete details on how to accelerate data processing with in-memory database technology, and also highlight how to accelerate the analysis of sequencing data by leveraging relevant information. With their work they eliminate time-consuming enquiries for relevant data (from disk storage) and enable instant interpretation of findings. This innovative approach should be of great value for applications ranging from research through to precision medicine.”

Scott Kahn, Illumina, CIO

“It will be essential to improve our understanding of the core functions of the human genome in order to develop stratified treatments for complex diseases and to provide a foundation for treatments to prevent or delay onset of diseases. By applying advanced in-memory technology to concrete problems of personalized medicine, Plattner and Schapranow demonstrate how interdisciplinary teams can develop innovative and appropriate solutions. Collaborative approaches of computational, scientific, and clinical teams have an enormous potential to improve the way we provide medical treatments in the future. Finally, the authors describe novel methods for flexible real-time analysis of medically relevant data that provide a powerful basis for timely decision making in personalized medical contexts.”

Prof. Dr. Peter N. Robinson, Charité, Head of the Computational Biology Group

“At Cytolon, we provide IT services to identify the most appropriate cord blood sample to limit patient's immune response. For this service, we need to analyze thousands of samples and combine them with a variety of heterogeneous patient properties. Plattner and Schapranow present that in-memory technology provides a meaningful way to integrate heterogeneous data. In addition, they show that real-time analyses of patient data is a paradigm shift in today's medicine. Thus, we believe this technology can help us to speed up the performance of our matching service.”

Thomas Klein, Cytolon AG, Founder and CEO

“At LGC Genomics, we build on our long-standing experience in providing DNA sequencing and analysis services to our customers. Latest sequencing machines have sped up extraction of DNA reads, but analysis is still time-intensive due to the sheer amount of generated data. Plattner and Schapranow apply the innovative in-memory technology to challenging analyses with impressive results. Long-running analysis processing, e.g. cohort analysis is reduced from taking up hours reduce to just a few seconds. We believe that this technology helps us to speed up our day-to-day business, allowing us to faster report back to our customer.”

Dr. Wolfgang Zimmermann, LGC Genomics, Business Unit Manager

“I am proud and thankful that HPI provides an environment that fosters teaching, research, and innovation in IT. Building on their former research results in database technology, Hasso Plattner and Matthieu Schapranow share insights of their high-performance in-memory genome platform that combines among others structured and unstructured medical data from various heterogeneous data sources to enable its real-time analyses in a single system. The platform is the outcome of a dedicated cooperation with various experts from biology, medicine, and computer science. As a result, it proves that interdisciplinary teams with actual knowledge from IT are able to considerably contribute in implementing the vision of great personalized medicine.”

Prof. Dr. Christoph Meinel, Hasso Plattner Institute, CEO

Preface

The human genome project was officially launched in 1990 equipped with a research funding of more than three billion USD. However, it took more than a decade and thousands of worldwide research institutes to discover and decode the full human genome sequence.

Nowadays, so-called next-generation sequencing devices process whole DNA and RNA within hours at moderate costs. Latest devices generate raw DNA reads with more than 30-times coverage in less than two days. However, interpretation and analysis of these raw data is still a time-consuming process potentially taking weeks. Next-generation sequencing devices are increasingly used in research and clinical environments to support treatment of specific diseases, such as cancer. This example highlights how fast the technological developments currently affect our daily lives.

Next-generation sequencing is also named to be the foundation for individual treatment decision, optimized therapies in course of personalized medicine and systems biology. Personalized medicine aims at treating patients specifically based on individual dispositions, such as genetic or environmental factors. However, the increasing amount of gathered diagnostic data requires specific software tools to identify relevant portions of data, process them at high-throughput, and provide ways to analyze them interactively.

We wrote this book to provide details about innovative approaches to process, combine, and analyze data required in the course of personalized treatment. It contains latest research results of applying in-memory database technology to process and analyze big genomic data. Furthermore, we share how to design and develop specific research tools that require real-time analysis of scientific data.

With this book, we contribute by bridging the gap between medical experts, such as physician, clinicians, and biological researchers, and technology experts, such as software developers, database specialists, and statisticians. As a result, we designed a specific structure of the book to support the individual audiences.

The book is structured as follows.

- Part I addresses the data acquisition, the modeling of processing and analysis pipelines, and how to accelerate preprocessing of data. This part is designed for bioinformaticians and researchers, who want to understand how to optimize the data preparation for their experiments.
- Part II gives examples how to design and implement specific applications enabling real-time analysis of scientific data. Furthermore, it provides guidelines to operate and to exchange huge data at fast pace. This part is intended for researchers and medical experts, who require to work with big data on a daily basis. It also provides guidelines for IT experts how to operate on these data from a software engineering perspective.

Potsdam, Oct 20, 2013

Hasso Plattner and Matthieu-P. Schapranow

Contents

Part I Data Processing in Personalized Medicine

Part II Real-Time Data Analysis in Personalized Medicine

List of Figures

List of Tables

List of Listings

Chapter 1
Innovations for Personalized Medicine

Hasso Plattner, Matthieu-P. Schapranow and Franziska Häger

The Human Genome (HG) project, officially launched in 1990, involved thousands of worldwide research institutes and required more than a decade to sequence and decode the full HG [5]. Next-Generation Sequencing (NGS) devices enable processing of whole genome data within hours while reducing costs [2]. NGS is used to support personalized medicine, which aims at treating patients specifically based on individual dispositions, e.g. genetic or environmental factors [20].

The In-Memory Database (IMDB) technology has proven to have major capabilities for analyzing big enterprise and medical data, e.g. to identify relevant patient data and to protect markets from injecting pharmaceutical counterfeits [41, 39].

In this work, we present our findings of applying IMDB technology to enable real-time analysis of genome data in course of our High-performance In-memory Genome (HIG) platform. We developed a specific IT platform that combines processing and analyzing of genomic data as a holistic process based on the feedback of researchers and clinicians. Our HIG architecture is designed to run on commodity hardware instead of highly specialized hardware to be a) cost-efficient and b) to make use of existing hardware infrastructures. Figure 1.9 depicts the system architecture of our HIG system modeled as block diagram using the Fundamental Modeling Concepts (FMC) [21].

In the following, we share requirements for IT systems in the course of personalized medicine that we acquired during our work in interdisciplinary teams. We outline latest changes in hardware that enable real-time analysis of big data with the help of in-memory database technology. With its help, we share details about our HIG system architecture and give an overview of the structure of this work.

1.1 Requirements for Personalized Medicine

Personalized medicine aims at treating patients specifically based on individual dispositions, e.g. genetic or environmental factors [20]. For that, researchers and physicians require a holistic view on all relevant patient specifics when making treatment decisions. Thus, the detailed acquisition of medical data is the foundation for personalized therapy decisions. The more fine-grained data are available, the more specific are the gained insights, but the complexity of data processing rises, too. This requires tool support to identify the relevant portion of data out of the increasing amount of acquired diagnostic data [41].

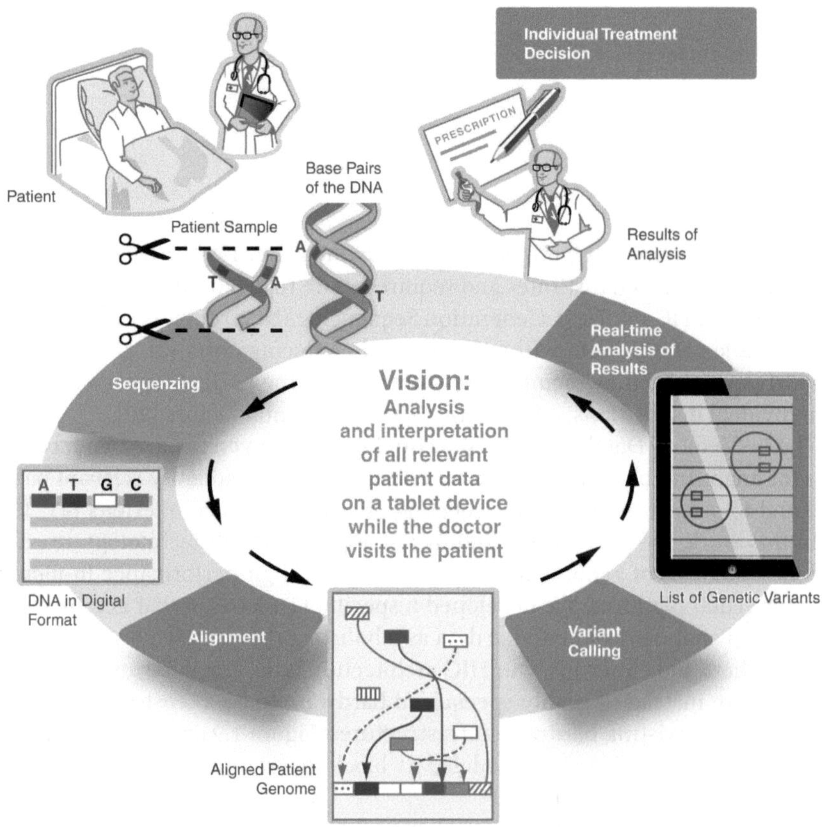

Fig. 1.1: Data processing steps involved in the analysis of genome data. Sequencing the samples results in chunks of DNA are available in digital form. During alignment their position within the whole genome is mapped. Variant calling results in a list of differences of a fixed reference. The analysis obtains new insights based on the list of detected variants.

Figure 1.1 depicts data processing steps required to include genome data in the course of personalized medicine. After a sample has been extracted, it is sequencing, which results in short chunks of DNA in digital form. The DNA chunks need to be aligned to reconstruct the whole genome and variants compared to a reference, e.g. normal vs. pathologic tissue, are detected during variant calling. The analysis of genome data builds on the list of detected variants, e.g. to identify driver mutations for a medical finding.

Personalized medicine requires solving clinical and technical issues before it can become a de facto standard in modern healthcare systems. For that, we identified the following end user roles in our conducted user research:

- **Researchers** work in clinical and pharmaceutical environments to acquire new knowledge, e.g. about therapies and pharmaceuticals,
- **Clinicians** have direct contact with patients in course of an actual treatment or therapy, and
- **Patients** suffer from either an actual or a chronic disease and want to recover in the most efficient way, e.g. in a short period of time.

In the following, we define specific requirements for designing clinical software systems. They reflect a selected subset of requirements from the software engineering catalog as defined for product quality in ISO/IEC 9126-1 and specifically revised in context of our work [19].

- **Ease of Use**: Clinical software artifacts must be usable by untrained users, i.e. its User Interface (UI) should combine ease of use and functionality.
- **Response Time**: The response time of clinical applications must not exceed an empirical threshold of approx. two second [11]. Our user interviews showed that otherwise the latency outperforms any benefits resulting in the application not being used.
- **Reliability**: Clinical software must be available without unplanned interruptions or malfunctions due to its life-critical purpose.
- **Productivity**: Users of a clinical software solution should be more efficient than performing manual processing steps or using alternative tools.
- **Scalability**: The system behavior of the designed software must not be affected by the number of concurrent users. Extending existing hardware resources, e.g. number of database servers, should result in a linearly increasing capacity.
- **Data Security**: Clinical data are sensitive and must be accessible by authorized personnel only. Intended or unintended exposure of these data must be addressed during the design of clinical software.

1.1.1 Researchers

Researchers work on discovering new medical knowledge, e.g. therapies, pharmaceuticals or influencing factors for certain diseases. They create hypotheses and need to validate them by specific experiments, e.g. in wet laboratories as depicted in Figure 1.2. For that, they acquire more and more experiment data that needs to be processed and analyzed, e.g. to identify correlation. Today researchers are often incorporating traditional office tools to analyze their data but find they cannot cope with the ever increasing amount of data they want to correlate. Therefore,

Fig. 1.2: Researchers work in clinical and pharmaceutical environments to acquire new knowledge, e.g. about therapies and pharmaceuticals.

analyses take hours or even days to be performed. This delay between stating a concrete hypothesis and getting its validation or falsification is essential. If this period of time is longer than a few minutes, the stream of thinking is disrupted. We found out that analyses need to be performed in an interactive way to improve efficiency of researcher's tasks. Thus, researchers need to involve IT experts and/or acquire specialized IT knowledge to optimize their data and perform appropriate analyses.

One popular example in the course of personalized medicine is genome data. Before they can be analyzed, time-consuming preprocessing needs to be performed as depicted in Figure 1.1. So the preprocessing can be considered as a batch job, while the analysis should be a real-time application. After starting the analysis the incorporated tools should provide instant results, e.g. show appropriate gene loci in a browser, identify relevant annotations, or list known associated diseases for a variant. Currently, any of these details are publicly available, but in different knowledge databases provided by individual institutes with dedicated search tools. These media breaks also result in a significant delay to check for certain details. Combining information from various sources and providing appropriate links in an automatic manner reduces media breaks and search time in individual databases. Further details about a concrete application example can be found in Section 1.5.1.

Another examples are biological pathways, which provide highly curated knowledge about cellular interactions in a graph format [13, Chap. 23]. Pathways are accessed as figures showing the graph or via search tools. However, the latter mainly check for containment relations, e.g. whether a certain gene or protein is present in a certain pathway, while the pathways' topology, e.g. whether certain genes are interconnected with each other or which distance they have, remains disregarded. Thus, the identification of appropriate pathways requires knowledge of their existence prior to search. With the list of variants, the search for appropriate pathways should be executed automatically. Relevant pathways need to be provided as a ranked list, e.g. genes having possible mutation sites and being interconnected within a certain pathway. Further details about a concrete application example can be found in Chapter 7.

Medical publications, such as paper or experiments, contain latest research results. However, finding relevant publications is still a manual job. It requires keyword search within international publication databases, such as PubMed [32]. For example, to take latest findings for a specific therapy approach into account, a constant search and continuing education is required. Furthermore, a variety of unstructured free-text documents are created in the course of patients' treatment, e.g. diagnoses, pathologic reports, and therapy decisions. The flexibility of a combined search in structured and unstructured medical data in a single system helps to find relevant details. For example, finding open clinical trials supports researchers to propose alternative therapy approaches. Further details about a concrete application example can be found in Chapter 8.

1.1.2 Clinicians

Fig. 1.3: Clinicians have direct contact to patients, e.g. in course of a actual treatment or therapy decisions. They want to find the optimal treatment decision for each individual patient out of a set of alternatives.

Clinicians have regularly contact to patients, e.g. during their visits as depicted in Figure 1.3. His or her aim is to find the most effective therapy for each patient. They need to decide on the next steps in course of the treatment for any individual patient. To take appropriate therapy decisions, clinicians are interested in evidence of appropriate therapies that have been successful in other cases. Thus, they are interested in identifying similar patient cases, e.g. to explore successful therapies or to prevent already documented side effects. Also in terms of preventive diagnostics, it becomes more and more relevant to identify subsequent diseases at the earliest possible stage. Analyzing similar cases helps to discover subsequent diseases or side effects before they are present in the current case. Identifying similar patient cases is a manual and time-consuming task. It requires that the clinician has this knowledge either from his/her own previous cases or from networking with colleagues about their former cases. Similar patient cases in other hospitals or countries are hard to identify today, e.g. due to a minimum of exchanged meta data from other clinics. With the help of appropriate patient data, such as gender, genetic variants, and preconditions, the classification of similar patients is possible. Furthermore, analysis tools can support the discovery of similar patient cases across hospital borders by exchanging pseudonymized patient meta data, e.g. the patient's history. With the help of cohort analysis it is possible to identify similar patient cases and to identify also patient cases that belong to another category, e.g. disease subtype. Further details about a concrete application example for secured exchange of medical data can be found in Chapter 4 and for patient cohort analysis in Chapter 6.

1.1.3 Patients

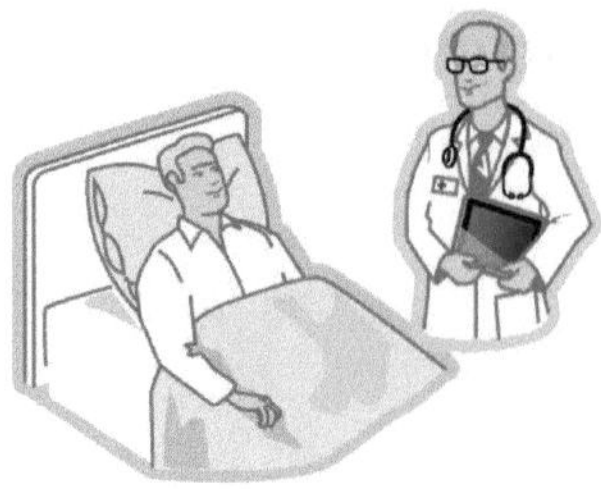

Fig. 1.4: Patients suffering from an actual or a chronic disease want to recover in the most efficient way.

Patients get in contact with clinicians if they suffer a concrete disease, e.g. being in hospital as depicted in Figure 1.4. Their chronic or actual diseases need to be addressed by individual therapies. For patients, it is important to recover in the most efficient way, e.g. by identifying appropriate pharmaceuticals or therapies. They also want to find the expert for their disease that maybe rare or complex so that they receive those therapies based on the latest knowledge of their disease. The identification of experts for a certain disease is hard even for physicians or medical experts. Thus, it is a possibility to identify experts by searching for clinical trials focusing on a certain disease or therapy. Another source for latest information is latest research literature, such as publications. Further details about a concrete application example can be found in Chapter 8.

1.2 Interdisciplinary Teams

During our work, we applied the Design Thinking (DT) methodology. The main idea of DT is to stay in close exchange with stakeholders and target users during the development of a solution. Thus, ensuring the final solution is desirable, viable, and feasible. A solution in that case could be anything that solves a problem or addresses a certain user need, e.g. a service, a product or a software application. A desirable solution will be helpful for end users and therefore will be wanted. A viable solution will be successful in the market, e.g. because cost performance ratio is considered good by end users. A feasible solution can be created with the current state of technology and knowledge available in the development team. Figure 1.5 depicts the relations between these three attributes.

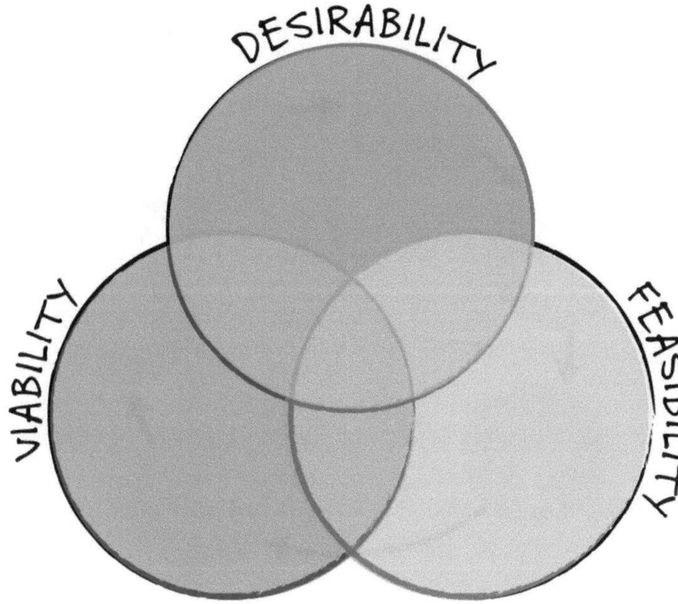

Fig. 1.5: Combined requirements of a problem solution: viability, desirability, and feasibility.

An important aspect of the DT methodology is the team composition. DT proposes to work in interdisciplinary teams to incorporate different views on the problem into the solution [35, Chap. 2.1]. The idea behind this proposal is that team members from different disciplines, e.g. a software developer and a medical researcher, will have different viewpoints on the same problem domain. Thus, if a team is comprised of members from different disciplines relevant to the problem at hand, chances that an important aspect is forgotten are minimized. Additionally, an interdisciplinary team will not suffer from rivalry between experts of the same field, instead all expertise necessary to implement the solution is already available in the team. Besides suggesting interdisciplinary team compositions, DT provides a process framework as depicted in Figure 1.6. It asks for constant communication between the developing team and the stakeholders and targeted end users. The methodology incorporates a variety of tools and methods that enable the development team to:

- Gain and collect information,
- Discover user needs,
- Identify hidden aspects in this information, and
- Communicate their ideas in a tangible manner.

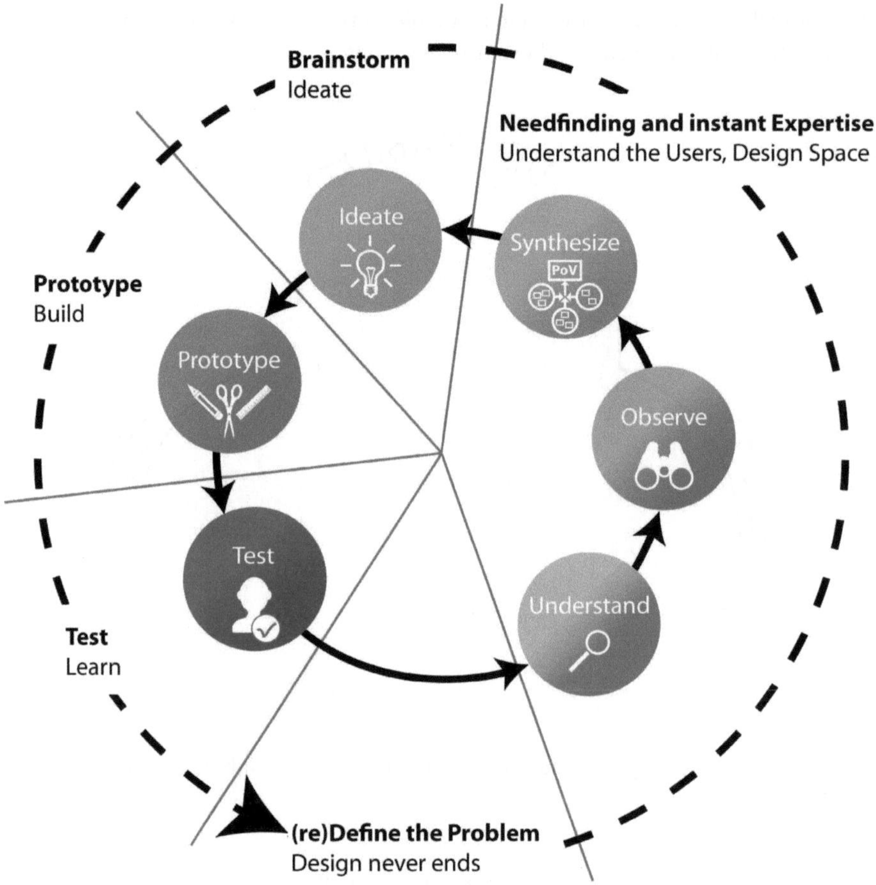

Fig. 1.6: Design thinking process as defined by the HPI School of Design Thinking in Potsdam and Stanford adopted from [28].

A DT team usually starts its process with an initial need finding phase, collecting information about target users, stakeholders, competitors and possible solutions, and gathering expertise. This can be done with a 360 degree research in literature, on the Internet, and a comparison of existing solutions in the problem domain, e.g. existing solutions in the field of genome analysis, medical databases and applications on the Internet, thus understanding the project, its goals, constraint and the environment. After this initial research, the team interviews and observes users and stakeholders identified in the first phase, e.g. physicians, researchers, and bioinformaticians.

All the information gathered in these two phases is synthesized into the team's point of view on the problem by clustering the available information and deriving so-called personas. A persona combines the most important aspects and insights from different interviews and serves as a representative of the target group to the development team, e.g. personas for the target groups researchers, clinicians and patients, as described in Section 1.1. Based on these personas and their respective needs and problems the team ideates on different aspects of a possible solution and creates prototypes of these ideas in a manner that focuses on transporting the main idea.

Prototypes are a central aspect of DT, as they allow users to experience a proposed solution or compare different solutions in a tangible manner and not just as an abstract description of an idea. Prototype fidelity can range from low-resolution prototypes like sketches or paper models to high-resolution prototypes like functioning miniature models or fully functional parts of the solution. They can not only be used to validate an idea or solution but can also help to learn and understand the user or certain aspects of the problem, e.g. by simulating certain situations and observe users in them. Each prototype should go through intensive testing with target users. The information gained by testing the ideas is synthesized again. Depending on the outcome of this synthesis phase, the team can start a consecutive iteration in which it moves on with further ideation to refine their ideas or, go back to understand and observe phases that answer open questions and investigate new aspects of the problem.

DT provides development teams with a simple process and an easy to use set of tools that ensures an iterative development in constant exchange with target users and stakeholders. During the process prototypes of the solution become more sophisticated and real until a final solution prototype or implementation of the ideas and features is reached.

1.3 Trends in Hardware

Computer hardware is continuously changing and improving. In recent years, multi-core architectures and larger main memory have been the most important trends. To keep up with the changes and exploit the hardware capabilities software systems, such as Clinical Decision Support Systems (CDSSs) must be constantly adapted. In this section, we will introduce recent hardware trends that enable real-time analysis of medical data.

Main Memory

A main driver for real-time analysis of large amounts of data is the development of IMDBs that are capable to process large volumes of data in a very fast response time as described in Section 1.4. Initial concepts for in-memory databases were created in the 1980s, but memory prices were too high and memory capacities too small for those systems to be viable for large applications [12]. Recent hardware development trends enabled further research of these concepts and the development of

IMDBs as a product for large application, e.g. enterprise software or CDSSs. Figure 1.7 compares prices of main memory, hard disk, and flash storage. It shows that costs have decreased exponentially over the past decades, leading to larger disk and main memory capacities in today's standard server systems. Besides storage costs

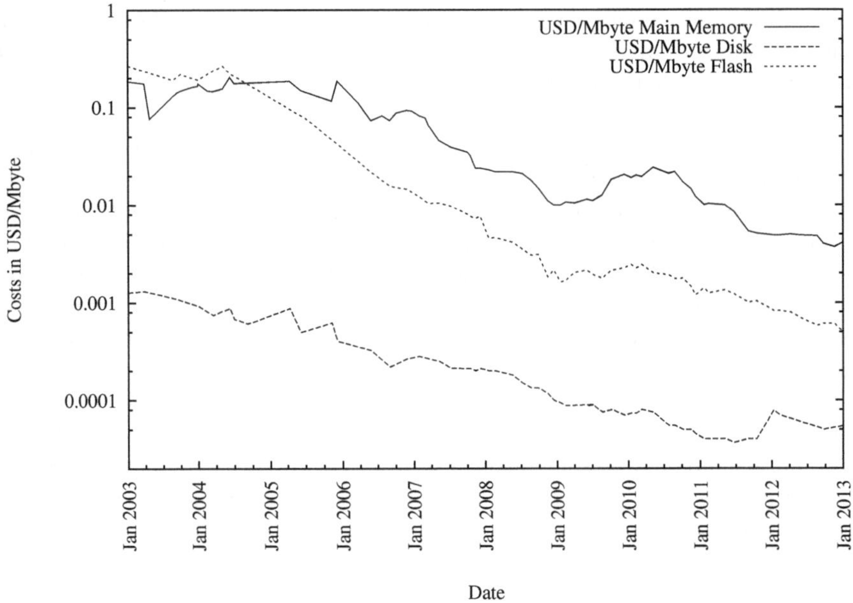

Fig. 1.7: Storage price development since 2003 based on data from [27].

and capacities, access speed is a very important enabler for real time analytics. It is also one of the main reasons why disk-based relational databases do not perform well when executing ad-hoc analytical queries over large data volumes. Simply accessing and reading the data from disk can take a significant amount of time, while in comparison the access speed of main memory is four orders of magnitude faster. As can be seen in Table 1.1, a main memory reference takes 100 ns, while current disks provide read and write seek times of about 5 ms [50, 16]. Thus main memory allows faster access without the need of providing special algorithms optimized for disk access.

Processor Development Trends

In 1965, Intel co-founder Gordon E. Moore made his famous prediction about the increasing complexity of integrated circuits in the semiconductor industry, also known as Moore's Law [30]. He stated that the number of transistors on a single

Action	Time [ns]
Main memory reference	100
Send 2,000 bytes over 1 Gb/s network	20,000
SSD random read	150,000
Read 1 MB sequentially from memory	250,000
Disk seek	10,000,000
Roundtrip time California to Germany	150,000,000

Table 1.1: Latency numbers for memory, disk and network taken from [34].

chip is doubled approximately every two years [18]. In reality the performance of Central Processing Units (CPUs) doubles in a 20-month timeframe on average. This is not only due to the increased number of transistors but also due to faster transistors, and more efficient circuitry. Performance and CPU frequency have stagnated after 30 years of exponential growth, because the circuit production has reached physical limits, e.g. the speed of light, power consumption, heat distribution [31]. To overcome this limitation, chip manufacturers such as Intel or AMD introduced multi-core processors. For high-end servers the first multi-core processors where introduced in 2001, followed by Intel's hyper-threading technology in 2002 that allowed better utilization of single- as well as multi-core processors by providing parallelism on a single core [3]. In 2005, multi-core processors were introduced to the consumer market. Kozyrakis et al. describe that the number of cores per chip used in home and business computers as well as high-end servers increased constantly since then, allowing the number of transistors per CPU to further increase according to Moore's Law [22]. Due to this development, software programs can no longer automatically benefit from the advances in processor technology. Instead, the developers of current and future software systems have to explicitly incorporate parallelism by splitting up algorithms across computing units or executing different operations concurrently.

Connection between Processor and Main Memory

The Front Side Bus (FSB) interconnects the CPU, main memory and other input/output components. The performance of the FSB developed much like the performance of CPUs. Analogous to single-core CPU performance, growth has stagnated. With the development of multi-core processors the FSB became the bottleneck of computers not being able to provide data from main memory as fast as multi-core processor could process the data. While compression techniques and specialized algorithms help to compensate this bottleneck for a small number of cores, the bottleneck becomes apparent for systems with many more cores (see Section 1.4). To overcome this problem chip manufacturers like Intel or AMD developed new technologies that integrate a direct connection from processor cores to main memory. With Intel's Quick Path Interconnect (QPI) and AMD's Hyper

Transport Protocol each processor core has local memory, that is adjacent to its own slot, and remote memory that is adjacent to other cores [1]. Computer architectures build on these technologies are called Non-Uniform Memory Access (NUMA) architectures. Data access to remote memory is up to 25 percent slower than access to local memory [10]. Thus, data locality is an important aspect for software applications and algorithms to consider. Parallel jobs need the data to be distributed across memory for each core to be able to work locally and algorithms need to ensure that most data can be read from local memory. Apart from the use of NUMA architectures in multi-core machines the architectural approach can also be used to consolidate multiple physical machines into one virtual machine. Note the difference in the commonly used term virtual machine. This term usually describes a setup where multiple parts of a physical machine are provided as multiple virtual machines, while in a NUMA setup multiple physical machines are bundled to give a user the impression of working with one extensively large server.

Next-generation Sequencing

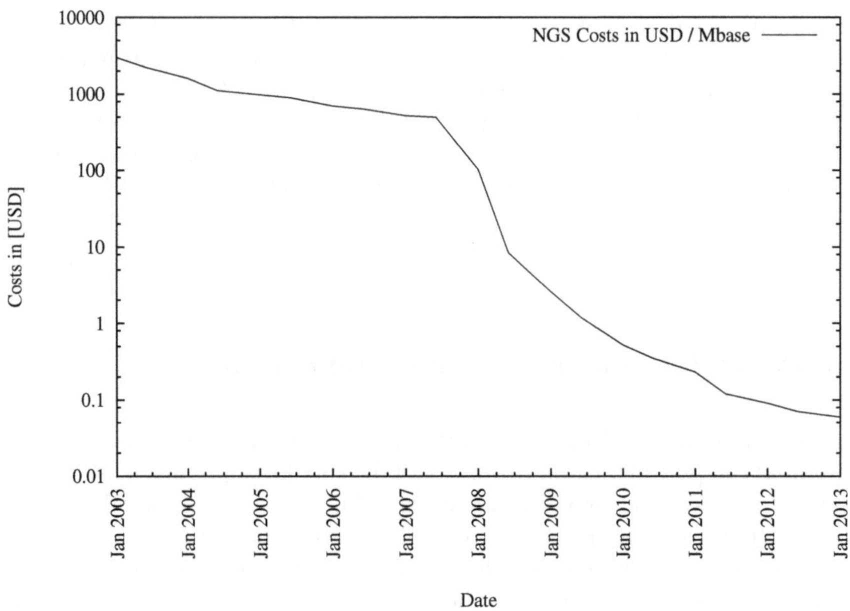

Fig. 1.8: Development of costs for next-generation sequencing between 2003 and 2013 adapted from [51].

Francis H. C. Crick and James D. Watson discovered in 1953 that the Deoxyribonucleic Acid (DNA) is built from a double helix structure [49]. Their discovery was based on the research results of Maurice H. Wilkins and Rosalind Franklin. In 1958, Rosalind died from Ovarian cancer at the age of 37. Crick, Wilkins, and Watson were honored in 1962 with the Nobel prize for medicine "for their discoveries concerning the molecular structure of nucleic acids and its significance for information transfer in living material" [33]. In the 1970s, several methods for manual sequencing of DNA, which is determining the order of nucleotides in the DNA's structure, were developed, e.g. by Sanger [26, 37].

In order to process whole genomes, such as the human genome, automated processes were needed. Leroy Hood from the California Institute of Technology and Michael Hunkapiler from Applied Biosystems, Inc. started to experiment with fluorescent dyes attached to a primer used in sequencing reactions in 1985. Their work resulted in the ABI 370 sequencing device in 1987, which was capable to support automatic sequencing. After several incremental improvements, the processing time for the whole human genome declined to some years [43].

Sequencing devices based on the Sanger method were used during the complete sequencing of the human genome in course of the HG project. It was started in 1990 and took more than a decade to come up with a first draft of the complete human genome. It involved thousands of research institutes worldwide for this fundamental work.

So-called Next-Generation Sequencing (NGS) devices accelerated the sequencing process dated back to the mid 1990s, e.g. devices from 454 Life Sciences [25].

Today these machines enable the processing of whole genome data within hours while reducing costs [2]. With latest sequencing technology, the costs for sequencing dropped rapidly to less than ten cents per raw megabase as depicted in Figure 1.8. As a result, more and more NGS devices are facilitated for research and clinical use. In the near future the adaption of NGS technology is expected to accelerate the complete sequencing process while maintaining moderate costs. This opens a variety of completely new diagnostic approaches building on the individual dispositions even on the genome level, e.g. in course of personalized medicine.

1.4 In-memory Technology Building Blocks

In the following, selected in-memory technology building blocks are introduced. We refer to in-memory technology as a toolbox of technological artifacts that enable processing of enterprise data in real-time in the main memory. This includes the processing of hundreds of thousands of queries in a multi-user system in sub-second response time. In-memory technology enables decision taking in an interactive way without keeping redundant or pre-aggregated data.

1.4.1 Combined Column and Row Store

To support analytical and transactional operations, two optimized types of database systems have evolved. On the one hand, database systems for transactional workloads store and process data in a row-oriented format, i.e. attributes are stored side by side. On the other hand, database systems optimized for analytical purposes scan selected attributes of huge datasets in a very short time, e.g. by maintaining pre-aggregated totals. If complete data of a single row needs to be accessed, storing data in a row format is advantageous. For example, the comparison of two customers involves all of their database attributes, such as inquirer's name, time, and content need to be loaded. In contrast, columnar stores benefit from their storage format, when only a subset of attributes needs to be processed. For example, summing up the gender ratio of patients treated in a certain period of time only involves the attributes date and gender, but the remainder, such as name and birth date, are not required. Using a row store for this purpose would require processing of all attributes, although only two of these attributes are required. Therefore, a columnar store benefits from accessing only relevant data.

1.4.2 Complete History

Keeping the history complete even after it has been updated or changed is the purpose of the insert-only or append-only technique to handle new and updated data. Traditional database systems support four operations for data manipulations, i.e. insert, select, delete, and update of data. The latter two are considered as destructive operations since original data are no longer available after its execution [38, Sect. 7.1]. In other words, it is neither possible to detect nor to reconstruct the complete history of values for a certain attribute after its execution since only the latest value is permanently stored. Insert-only tables enable storing the complete history of value changes and the latest value for a certain attribute [34]. For instance, this is the foundation of all bookkeeping systems to guarantee transparency. In addition, insert-only enables tracing of access decisions, e.g. to perform incident analysis.

1.4.3 Lightweight Compression

Compression in context of in-memory technology refers to a mapping that defines a storage representation, which consumes less space than the original representation [34, Chap. 7]. A columnar storage supports the use of lightweight compression techniques, such as run-length encoding, dictionary encoding, and differencing [44]. Due to the fixed data type per column, subsequent values are within a given interval, e.g. integer values. In addition, the given data type defines an upper threshold for individual values. Depending on the source of data, the number of concrete values is lower than all possible values from this interval, i.e. the amount of distinct values is required to store all data. This data representation requires only the amount of distinct values to store. For example, the International Code of Diseases (ICD) is identical for patients suffering from the same disease. Instead of storing the ICD redundantly, dictionary compression stores it once and maps it to a smaller integer representation. Thus, only the corresponding integer value is stored in the database and all queries are rewritten to use the integer representation instead. The original representation is replaced just before the result set is returned to the client. As a result, the database executes all operations on compressed data without decompression. In comparison to the uncompressed format, the compressed data improves cache-hit ratio since more compressed data fits into the same amount of cache memory.

1.4.4 Partitioning

We distinguish between vertical and horizontal partitioning approaches [24]. Vertical partitioning refers to rearranging individual database columns. It is achieved by splitting columns of a database table in two or more sets of columns. Each of the sets can be distributed on individual databases servers. This technique can also be used to build up database columns with different ordering to achieve better search performance while guaranteeing high-availability of data [15]. Key to success of vertical partitioning is a thorough understanding of data access patterns. Attributes that are accessed in the same query should rely in the same partition since locating and joining additional columns result in degradation of overall performance.

In contrast, horizontal partitioning addresses long database tables and how to divide them into smaller chunks of data. As a result, each portion of the database table contains a subset of the complete data. Splitting data in equivalent long horizontal partitions is used to support parallel search operations and to improve scalability. For example, the identification of patients with certain preconditions requires a complete scan of the corresponding database columns. With a single partition, a single thread needs to access all individual history entries to check the relevant predicate for selection. When using a naive round robin horizontal partitioning across ten partitions, the scan of the complete table is performed in parallel by ten threads simultaneously. It reduces the response time by approx. one ninth compared to the aforementioned single threaded approach.

This example shows that the partition length depends on the incorporated partitioning strategy. For example, instead of using round robin, range partitioning can be used, e.g. patient data are grouped according to their last treatment year.

1.4.5 Multi-core and Parallelization

Parallelization can be applied to various locations within the application stack of enterprise systems – from within the application running on an application server to query execution in the database system. As an example of application-level parallelism, we assume the following. Multiple clinical departments access the data of a single patient simultaneously. Processing multiple queries can be handled by multi-threaded applications, i.e. the application does not stall when dealing with more than one query at a time. Operating systems threads are a software abstraction that needs to be mapped to physically available hardware resources [45, Chap. 2]. A CPU core is comparable to a single worker on a construction area. If it is possible to map each query to a single core, the system's response time is optimal. Query processing also involves data parallelization, i.e. the database needs to be queried in parallel, too. If the database is able to distribute the workload across multiple cores a single server works optimal. If the workload exceeds physical capacities of a single system, multiple servers or blades need to be installed for distribution of work to achieve optimal processing behavior. From the database point of view, data partitioning supports parallelization since multiple CPU cores even on multiple servers can process data simultaneously [14, Chap. 6]. This example shows that multi-core architectures and parallelization depend on each other while partitioning forms the basis for parallel data processing.

1.4.6 Active and Passive Data Store

We define two categories of data stores: active and passive. We refer to active data when it is accessed frequently and regular updates are expected, e.g. data of patients currently treated in a hospital. In contrast, passive data are neither updated nor accessed regularly. They are purely used for analytical and statistical purposes or in exceptional situations where specific investigations require these data. For example, tracking events of a certain pharmaceutical product that was sold five years ago can be considered as passive data. Firstly, from the business' perspective, the pharmaceutical can be consumed until the best-before date of two years after its manufacturing date is reached. When the product is handled now, five years after its manufacturing, it is no longer allowed to be sold. Secondly, the product was most probably sold to a customer four years ago, i.e. it left the supply chain and is typically already used within its best-before data. Therefore, the probability that details about this pharmaceutical are queried is very low. Nonetheless, the tracking history is conserved and no data is deleted in conformance to legal regulations. As a result, the passive data can still be accessed but with a higher latency than active data. For example, passive data can be used for reconstructing the path of a product within the supply chain or for a financial long-term forecast. This example gives an understanding about active and passive data. Furthermore, introducing the concept of passive data results in a classification of data stores. Thus, active data that needs to be accessed in real-time can be separated from passive data that is ready for archiving.

When data are moved to a passive data store, they no longer consume fast accessible main memory. Dealing with passive data stores involves the definition of a memory hierarchy including fast, but expensive, and slow, but cheap memory. A possible storage hierarchy is given by: memory registers, cache memory, main memory, flash storages, solid state disks, SAS hard disk drives, SATA hard disk drives, tapes, etc. To distinguish between active and passive data, rules for migration of data from one store to another needs to be defined. We refer to them as data aging strategy or aging rules. We consider the process of aging, i.e. the migration of data from a fast to a slower store as background task, which is performed regularly, e.g. once a month or once a week. Since this process involves reorganization of the entire database, it should be performed only during times of low database access, e.g. at night or on weekends.

1.4.7 Reduction of Layers

In application development, layers refer to levels of abstraction. Each application layer encapsulates specific logic and offers certain functionality. Although abstraction helps to reduce complexity, it also introduces obstacles. Examples for the latter are a) functionality is hidden within a layer and b) each layer offers a variety of functionality while only a small subset is in use. From the data point of view, layers are problematic since data are marshaled and un-marshaled for transformation to the layer-specific format. As a result, identical data are present in various layers and representations. A reduction of layers improves the use of hardware resources. Moving application logic to the data it operates on results in a smaller application stack and also code reduction. Finally, reducing the code length results in improved maintainability.

1.5 High-performance In-memory Genome Platform

We developed the High-performance In-memory Genome (HIG) platform to conduct research on it[1]. This dedicated platform provides a framework for execution of specific micro applications to answer specific research questions [40]. The system architecture is modeled in Figure 1.9 and consists of the layers: application, platform, and data, which are described in detail in the following.

[1] http://we.analyzegenomes.com/

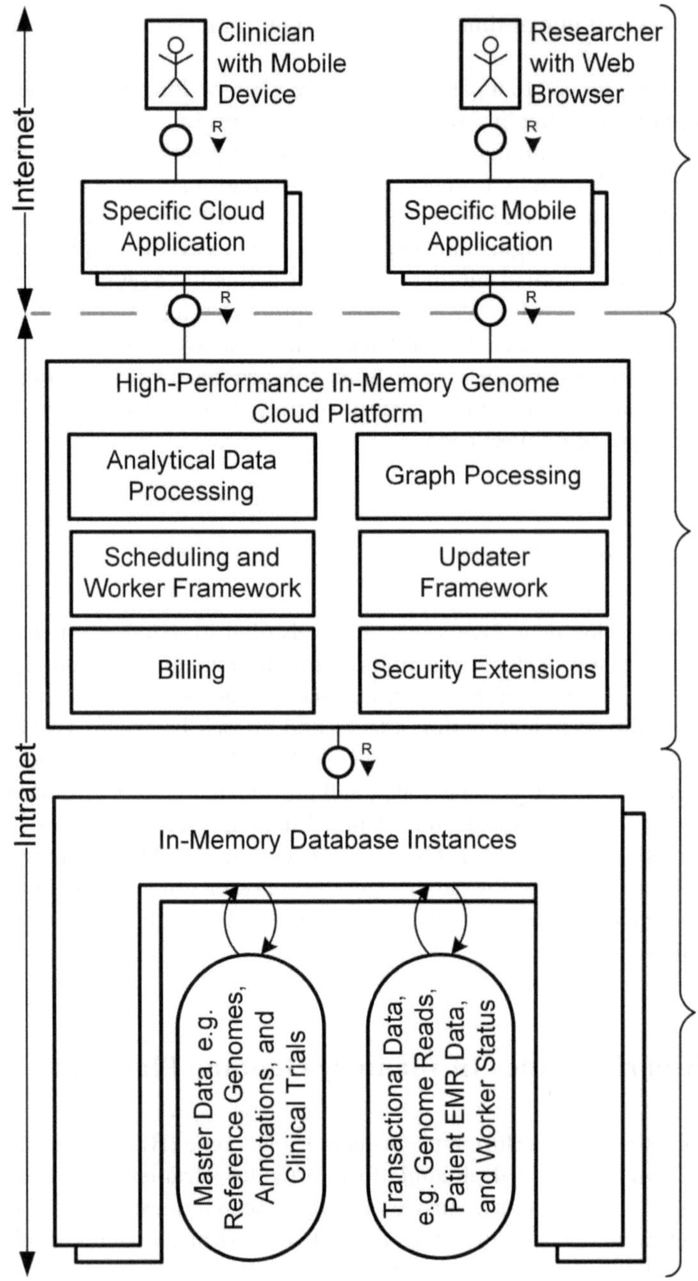

Fig. 1.9: The HIG system architecture consists of application, platform, and data layer. Analysis and processing of data are performed within the platform layer eliminating time-consuming data transfer.

1.5.1 Application Layer with Micro Applications

The application layer consists of special purpose applications to answer medical and research questions. We provide an Application Programming Interface (API) that can be consumed by various kinds of applications, such as web browser applications or mobile applications. Figure 1.9 depicts the data exchange via asynchronous Ajax calls and JavaScript Object Notation (JSON) [17, 6]. As a result, performing specific analyses is no longer limited to a specific location, e.g. the desktop computer of a clinician. Instead, all applications can be accessed via devices connected to the Internet, e.g. laptop, mobile phone, or tablet computer. This enhances the user's productivity by having access to relevant data at any time.

In the following, we outline selected micro applications and how they are combined to implement an end-to-end process for personalized medicine.

Alignment Coordinator

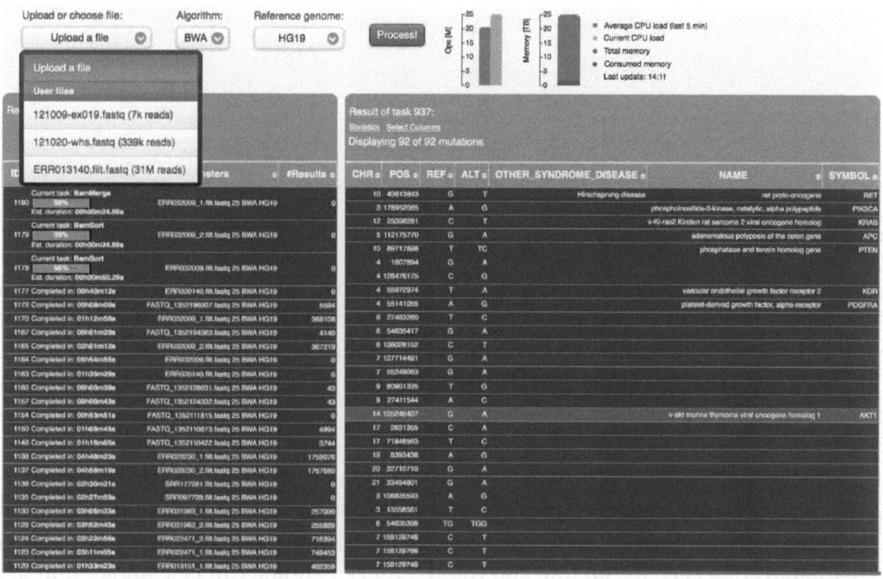

Fig. 1.10: HIG Alignment Coordinator: From top to bottom, configure parameters for data processing, status of recent tasks on the left, and investigate result sets of a selected task on the right.

The alignment coordinator cockpit as depicted in Figure 1.10 is used to issue new, supervise existing, and retrieve the results of completed runs of genome data processing. The input is a FASTQ file, a selected pipeline configuration consisting of specific alignment algorithms and variant calling steps, and the reference genome to use, as well as, pipeline specific parameters.

Selecting an entry from the tasks lists on the left displays the results of the variant calling in a table on the right. Attributes of the results table can be configured individually, e.g. associated diseases, affected genes, or similar cases, can be included in the table. By clicking on a certain mutation, the specific chromosome location is displayed in detail using the genome browser application.

Genome Browser

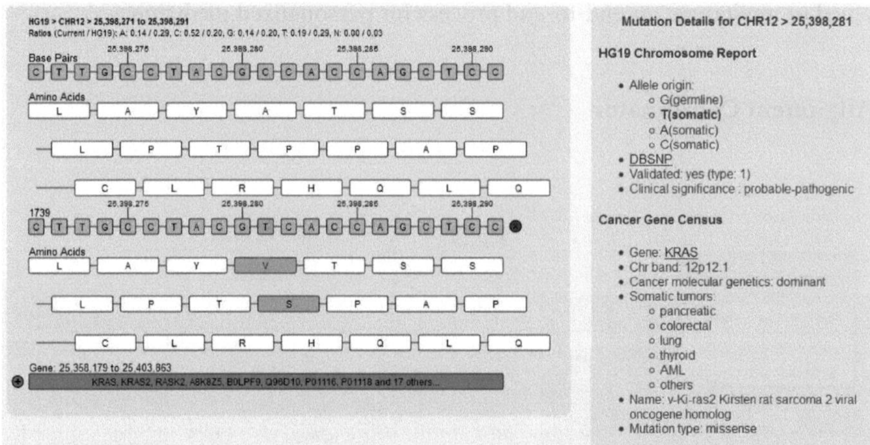

Fig. 1.11: HIG Browser: From top to bottom, comparison of reference genome and a selected cell lines at a specific locus on chromosome 12. A SNP is depicted on the left. Combined information from international research databases about the selected mutation are shown on the right.

Figure 1.11 depicts a screenshot of our genome browser. It is a specific application, which enables analysis of a specific mutation of certain cell lines or patient genomes with each other. Base pairs, amino acids, gene splicing variants, and further details of a certain cell line and the reference genome can be compared in detail. Thus, the cellular impact of mutations can be investigated and the excerpt of the DNA can be compared between various patients using this micro application. Known variants are highlighted by automatically combining worldwide annotation databases, and relevant details and links to existing database, such as dbSNP, DGV, Sanger, are displayed when a certain mutation is selected [9, 42, 46].

Cohort Analysis

The HIG cohort analysis micro application enables researchers and clinicians to perform interactive clustering on the data stored in the IMDB. For example, it enables k-means and hierarchical clustering on selected sets of data [23, Chap. 13]. Thus, users are able to verify hypotheses by combining patient and genome data in real-time.

Clinical Trial Search

The HIG clinical trial search micro application assists physicians in finding adequate clinical trials for their patients. It analyses patient data, such as age, gender, preconditions and existing mutations, and matches them with open clinical trials. The analysis is performed on top of more than 30k descriptions of recruiting clinical trials, which are ranked in real-time while the physician investigates the list of variants in the patient's genome [48].

1.5.2 Platform Layer

The platform layer holds the complete process logic and consists of the IMDB system for enabling real-time analysis of genomic data. We developed specific extensions for the IMDB system to support genome data processing and its analysis. In the following, selected extensions and their integration in the IMDB system are described in greater detail.

Scheduling of Data Processing

We extended the IMDB by a worker framework, which executes tasks asynchronously, e.g. alignment of chunks of genome data. It consists of a task scheduler instance and a number of workers controlling dedicated computing resources, e.g. individual computing nodes. The scheduler is contacted by the web service, i.e. the scheduler is the central coordinator for asynchronous job execution. Workers retrieve tasks and parameters by the scheduler instance and perform specific tasks, such as workbench preparation, task execution, and maintenance of status information. Thus, all workers are connected to the IMDB to store status information about currently executed tasks.

Furthermore, the scheduler supervises the responsiveness of individual compute resources. If a predefined response behavior is no longer guaranteed, e.g. due to an overloaded compute node or a crashed worker process, workers are marked as unresponsive. As a result, work-in-progress tasks of the unresponsive worker are reassigned to a new worker to guarantee their execution and the unresponsive worker is scheduled for a restart or the administrator is informed.

Updater Framework

We consider the use of latest international research results as an enabler for evidence-based therapy decision [40]. The updater framework is the basis for combining international research results. It periodically checks all registered Internet sources, such as public FTP servers or web sites, for updated and newly added versions of annotations, e.g. database exports as dumps or characteristic file formats, such as CSV, TSV, and VCF. If the online version is newer than the local available version, the new data are automatically downloaded and imported in the IMDB to extend the knowledge base.

The import of new versions of research databases is performed as a background job without affecting the system's operation. We import new data without any data transformations in advance. Thus, they are available for real-time analysis without any delay [4, 8]. For example, the following selected research databases are checked regularly by our updater framework: National Center for Biotechnology Information (NCBI), Sanger's catalogue of somatic mutations in cancer, University of California, Santa Cruz (UCSC) [52, 9, 29].

1.5.3 Data Layer

The data layer holds all required data for processing and analyzing of genomic data. The data can be distinguished into two categories: master data and transactional data [7]. Human reference genomes, genome annotation data, and clinical trials data are referred to as master data, whereas patient-specific NGS data, Electronic Medical Records (EMRs), and the current system status are referred to as transactional data [47, 36]. Its analysis is the basis for gathering specific insights, e.g. individual genetic dispositions, and for leveraging personalized treatment decision in course of personalized medicine [20].

The actual step of analyzing the genetic data requires very specific questions to be answered. The application layer consists of specific applications for that purpose. They make use of the platform layer to initialize the data processing.

1.6 Structure of the Work

The following section gives you an overview of the book and its content. Table 1.2 provides an easy navigation to identify relevant topics for the different target groups of the book.

Title	Chapter	Audience
Modeling Genome Data Processing Pipelines	2	B, R
Scheduling and Execution of Genome Data Processing Pipelines	3	B
Exchanging Medical Knowledge	4	B, R
Billing Processes in Personalized Medicine	5	C, R
Real-time Analysis of Patient Cohorts	6	C, R
Ad-hoc Analysis of Genetic Pathways	7	C, R
Combined Search in Structured and Unstructured Medical Data	8	C, P, R
Real-time Collaboration in the Course of Personalized Medicine	9	B, C, P, R

Table 1.2: Mapping of chapters to addressed audience (B = Bioinformaticians, C = Clinicians, P = Patients, R = Researchers).

The remainder of the book is structured as follows. In Part I insights about data processing in course of personalized medicine are given. The target group for the following chapters is bioinformatics and researchers, who want to optimize data processing for analysis.

Chapter 2 presents details of how to model genome data processing pipelines using established business process modeling tools. It enables a systematic approach for creating models of genome data processing pipelines and builds the foundation for discussing and adapting concrete process model instances.

Chapter 3 discusses the execution of concrete genome data processing instances on top of an in-memory database. Furthermore, the parallel execution of individual pipeline tasks and concurrent processing of multiple requests to achieve high throughput are addressed.

The results of genome data preprocessing and the combination with relevant data build the basis for genome data analysis.

The exchange of medical knowledge in course of personalized medicine support the discovery of new knowledge, e.g. similar patient cases. Chapter 4 gives results about sharing data for research purposes without losing the ownership of the intellectual property.

Sharing medical knowledge is supported by an adequate compensation for the provided content. For that, real-time billing for personalized medicine is described in Chapter 5.

Part II addresses the real-time analysis of genome data, e.g. to discover evidence-based knowledge. The target group for the following chapters is medical experts and researchers who acquire big medical data in course of their daily work.

In Chapter 6 the real-time analysis of patient cohort data built on in-memory database technology is presented. The use of selected statistical clustering methods, such as k-means or hierarchical clustering, to form individual patient clusters based on genomic data is shared.

The interactive exploration of relevant genetic pathways is outlined in Chapter 7. Take the pathway's topology into account enables precise ranking of pathways based on the patient's individual genetic variants.

Chapter 8 explores the combined search in structured and unstructured medical data. For that, it shows how to identify recruiting clinical trials automatically by extracting relevant entities, such as preconditions, genetic variants, and pharmaceutical ingredients.

Ultimately, Chapter 9 outlines selected innovative applications and business processes that become available with real-time data explorations. It discusses where fiction is already reality and how the personalized medicine is enabled by instant data processing.

1.7 References

[1] Advanced Micro Devices I (2001) HyperTransport Technology I/O Link, A High-Bandwidth I/O Architecture

[2] Ansorge WJ (2009) Next-generation DNA Sequencing Techniques. New Biotechnology 25(4):195–203

[3] Behling S et al. (2001) The POWER4 Processor Introduction and Tuning Guide. http://www.redbooks.ibm.com/redbooks/pdfs/sg247041.pdf. Accessed Sep 23, 2013

[4] Bog A, Sachs K, Plattner H (2012) Interactive Performance Monitoring of a Composite OLTP and OLAP Workload. In: Proceedings of the International Conference on Management of Data, ACM, Scottsdale, AZ, USA, pp 645–648

[5] Collins FS et al. (1998) New Goals for the U.S. Human Genome Project. Science 282(5389):682–689

[6] Crockford D (2006) RFC4627: The application/json Media Type for JavaScript Object Notation (JSON). http://www.ietf.org/rfc/rfc4627.txt. Accessed Sep 23, 2013

[7] Das TK, Mishra MR (2011) A Study on Challenges and Opportunities in Master Data Management. International Journal of Database Management Systems 3(2)

[8] Färber F et al. (2012) SAP HANA Database: Data Management for Modern Business Applications. Sigmod Record 40(4):45–51

[9] Forbes SA et al. (2010) The Catalogue of Somatic Mutations in Cancer: A Resource to Investigate Acquired Mutations in Human Cancer. Nucl Acids Res 38

[10] Fujitsu Technology Solutions (2010) Memory Performance of Xeon 7500 (Nehalem-EX) Based Systems. http://globalsp.ts.fujitsu.com/dmsp/Publications/public/wp-nehalem-ex-memory-performance-ww-en.pdf. Accessed Sep 23, 2013

[11] Galitz W (2002) The Essential Guide to User Interface Design: An Introduction to GUI Design Principles and Techniques. John Wiley & Sons, New York, NY, USA

[12] Garcia-Molina H, Salem K (1992) Main Memory Database Systems: An Overview. IEEE Transactions on Knowledge and Data Engineering 4(6):509–516

[13] Ginsburg G, Willard H (2012) Genomic and Personalized Medicine. Elsevier Science

[14] Hellerstein JM, Stonebraker M (2005) Readings in Database Systems, 4th edn. MIT Press

[15] Hellerstein JM, Stonebraker M, Hamilton J (2007) Architecture of a Database System, Foundation and Trends in Databases, vol 1. now Publishers

[16] Hitachi (2010) Hitachi Global Storage Technologies, Ultrastar 15K450. http://www.hitachigst.com/internal-drives/enterprise/ultrastar/ultrastar-15k450. Accessed Sep 23, 2013
[17] Holdener AT (2008) AJAX: The Definitive Guide, 1st edn. O'Reilly
[18] Intel Corporation (2010) Microprocessor Quick Reference Guide. http://www.intel.com/pressroom/kits/quickrefyr.htm. Accessed Sep 23, 2013
[19] ISO (2001) 9126-1: Software Engineering: Product Quality
[20] Jain K (2009) Textbook of Personalized Medicine. Springer
[21] Knöpfel A, Grone B, Tabeling P (2006) Fundamental Modeling Concepts: Effective Communication of IT Systems. John Wiley & Sons
[22] Kozyrakis C et al. (2010) Server Engineering Insights for Large-Scale Online Services. IEEE Micro 30(4):8–19
[23] Krawetz S (2009) Bioinformatics for Systems Biology. Humana Press
[24] Lightstone S et al. (2007) Physical Database Design: The Database Professional's Guide to Exploiting Indexes, Views, Storage, and more. Morgan Kaufmann
[25] Margulies M et al. (2005) Genome Sequencing in Microfabricated High-density Picolitre Reactors. Nature 437(7057):376–380
[26] Maxam AM, Gilbert W (1977) A new method for sequencing DNA. Proceedings of the National Academy of Sciences 74(2):560–564
[27] McCallum JC (2013) Memory Prices (1957-2013). http://www.jcmit.com/memoryprice.htm. Accessed Sep 23, 2013
[28] Meinel C, Leifer L (2011) Design Thinking: Understand, Improve, Apply. Springer
[29] Meyer LR et al. (2012) The UCSC Genome Browser Database: Extensions and Updates 2013. Nucl Acids Res
[30] Moore G (1965) Cramming More Components onto Integrated Circuits. Electronics Magazine 38
[31] Naffziger S, Warnock J, Knapp H (2005) SE2 When Processors Hit the Power Wall (or "When the CPU Hits the Fan"). In: Proceedings of the International Solid-State Circuits Conference, San Francisco, CA, USA, pp 16–17
[32] National Center for Biotechnology Information, U.S. National Library of Medicine (2013) Pubmed. http://www.ncbi.nlm.nih.gov/pubmed. Accessed Sep 23, 2013
[33] Nobel Media AB (2013) The Nobel Prize in Physiology or Medicine 1962. http://www.nobelprize.org/nobel_prizes/medicine/laureates/1962. Accessed Sep 23, 2013
[34] Plattner H (2013) A Course in In-Memory Data Management: The Inner Mechanics of In-Memory Databases. Springer
[35] Plattner H, Meinel C, Weinberg U (2009) Design-Thinking. Mi-Fachverlag
[36] Rector A, Nolan W, Kay S (1991) Foundations for an Electronic Medical Record. Methods of Information in Medicine pp 179–186
[37] Sanger F, Nicklen S, Coulson AR (1977) DNA sequencing with chain-terminating inhibitors. Proceedings of the National Academy of Sciences 74(12):5463–5467

[38] Schapranow MP (2008) Transaction Processing 2.0. Master's thesis, Hasso Plattner Institute
[39] Schapranow MP (2013) Real-time Security Extensions for EPCglobal Networks: Case Study for the Pharmaceutical Industry. Springer
[40] Schapranow MP, Plattner H, Meinel C (2013) Applied In-Memory Technology for High-Throughput Genome Data Processing and Real-time Analysis. In: Proceedings of the XXI Winter Course of the Centro Avanzado Tecnológico de Análisis de Imagen, pp 35–42
[41] Schapranow MP et al. (2013) Mobile Real-time Analysis of Patient Data for Advanced Decision Support in Personalized Medicine. In: Proceedings of the 5th International Conference on eHealth, Telemedicine, and Social Medicine
[42] Sherry ST et al. (2001) dbSNP: The NCBI Database of Genetic Variation. Nucleic Acids Research 29(308–311)
[43] Stein RA (2008) Next-Generation Sequencing Update. Genetic Engineering & Biotechnology News 28(15)
[44] Svensson P (2008) The Evolution of Vertical Database Architectures – A Historical Review. In: Proceedings of the 20th International Conference on Scientific and Statistical Database Management, Springer-Verlag, pp 3–5
[45] Tanenbaum AS (2009) Modern Operating Systems, 3rd edn. Pearson Prentice Hall
[46] The Centre for Applied Genomics (2013) Database of Genomic Variants. http://dgvbeta.tcag.ca/dgv/app/downloads. Accessed Sep 23, 2013
[47] The Genome Reference Consortium (2013) Genome Assemblies. http://www.ncbi.nlm.nih.gov/projects/genome/assembly/grc/data.shtml. Accessed Sep 23, 2013
[48] U.S. National Institutes of Health (2013) ClinicalTrials.gov. http://www.clinicaltrials.gov/. Accessed Sep 23, 2013
[49] Watson JD et al. (1953) Molecular structure of nucleic acids. Nature 171(4356):737–738
[50] Western Digital Corporation (2010) Specification for the Seraial AT 6 Gb/s VelociRaptor Enterprise Hard Drives. http://wdc.custhelp.com/app/answers/detail/a_id/5377. Accessed Sep 23, 2013
[51] Wetterstrand K (2013) DNA Sequencing Costs: Data from the NHGRI Genome Sequencing Program (GSP). http://www.genome.gov/sequencingcosts/. Accessed Sep 23, 2013
[52] Wheeler DL et al. (2008) Database Resources of the National Center for Biotechnology Information. Nucleic Acids Research 36:D13–D21

Part I
Data Processing in Personalized Medicine

This part addresses data acquisition, modeling of processing and analysis pipelines, and how data preprocessing can be accelerated with the help of in-memory database technology. The content is designed for bioinformatics and researchers, who want to understand how to optimize the data preparation for their experiments.

Figure I.1 depicts the HIG system architecture as FMC block diagram focusing on IT components relevant for this part. For example, it shows the modeling engine for analysis pipeline, security extensions to ensure secured exchange of data, and the application runtime to integrate and execute specific analysis tools.

Table I.1 provides an easy navigation to identify relevant topics and target groups within this part of the book.

Title	Chapter	Audience
Modeling Genome Data Processing Pipelines	2	B, R
Scheduling and Execution of Genome Data Processing Pipelines	3	B
Exchanging Medical Knowledge	4	B, R
Billing Processes in Personalized Medicine	5	C, R

Table I.1: Structure of Part I and addressed audiences (B = Bioinformaticians, C = Clinicians, R = Researchers).

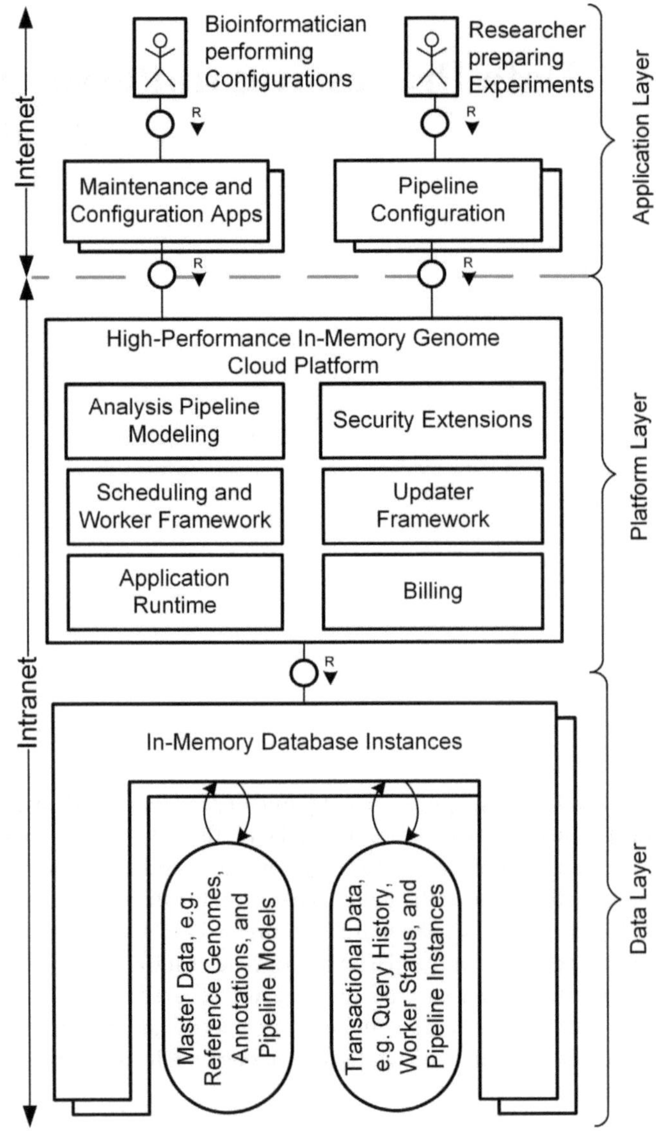

Fig. I.1: The HIG system architecture for data processing.

Chapter 2
Modeling Genome Data Processing Pipelines

Marie Schäffer

Abstract In order to conduct analyses on genome data, different calculation steps have to be done in a specific order, which constitutes a genome data processing pipeline. Still a lot of research is in process, in order to find faster and more reliable ways to do various analyses, so single steps or the whole sequence of the pipelines might be subject to change. A modular and flexible way to configure pipelines could simplify their use and the sharing of pipelines between researchers. With a possibility to configure pipelines without altering source code, bioinformaticians and technicians would be relieved of the task to rewrite a pipeline every time a single algorithm changes. This contribution proposes to use common process modeling tools for the abstract representation of genome data processing pipelines. The benefits and drawbacks of different process model notations are examined with special focus on the possibilities to specify execution semantics. As a prototype, a system for the parsing and execution of genome data processing pipelines specified in business process model and notation, is introduced.

2.1 Introduction

The analysis of genome data is an important element of both research and treatment of many diseases. Any advanced analyses consist of several steps that have to be executed in a specific order. The most fundamental of these analyses is the alignment, which assembles the genetic information from small DNA fragments. The succession of such analysis tasks can be represented as a pipeline and will be referenced as genome data processing pipeline henceforth.

Today, these pipelines often consist of a number of scripts calling each other directly. This approach has significant drawbacks: Every time the structure of a pipeline changes, a technician has to change the code of several jobs.

To achieve better maintainability, I propose a model-based approach for the definition of genome data processing pipelines. The objective is to deduce the succession of jobs from a graphical model of the pipeline. A scheduler that is responsible for the initiation and completion of jobs as described in Chapter 3 manages the execution. With such a system, every job would be stored once and get all pipeline specific parameters from the scheduler. Thus, algorithmic changes would only have to be incorporated in one script. Structural changes of pipelines could be incorporated without touching any source code.

Another goal of the approach described in this work is to enable external scientists and physicians to model their pipelines according to their individual needs and have them executed on a central computer cluster. I discuss the different requirements of the two user groups and introduce a corresponding research prototype.

The remainder of this work is structured as follows: In Section 2.2, different approaches for the modeling and execution of process pipelines are examined as well as other approaches concerning an easy interface for genome analysis processes.

Section 2.3 covers the steps needed to design executable pipeline models. First the general requirements for the representation of genome data processing pipelines are examined including the modeling of information flow and parameters for jobs. The different modeling notations are assessed for their applicability to the specified requirements. The degree of specification and standardization is taken into account as well as the question how intuitive user groups other than informaticians may use them. I describe one way to meet these requirements by using Business Process Model and Notation (BPMN) and applying an abstract view of the pipeline. The representation of such a BPMN model in the XML Process Definition Language (XPDL) is also discussed.

In Section 2.4, the research prototype is described in detail. I outline the desired functionality for programmers, researchers and physicians. The structure of the prototype is explained including the user interface, as this is an important factor influencing the benefit for the target group.

The evaluation in Section 2.5 focuses on the usability for the different target groups. Modeling decisions made for the prototype are reviewed and compared to alternative methods and the advantages and disadvantages of the approach used are compared to the systems discussed in Section 2.2.

In Section 2.6, I discuss the impact this approach could have and extensions that could be made to the basic system introduced here.

2.2 Related Work

A number of different notations for the creation of process models are used for different purposes. All of them have benefits as well as drawbacks for the desired usage. In this section, I introduce selected notations that might be fitting for the modeling of executable genome data processing pipelines.

Furthermore, I introduce approaches to the offering of a simple interface for genome analysis similar to my own.

Petri Nets

The concept of Petri nets is a system for the graphical modeling of complex process logic [65, Chap. 4.2]. It originates from Carl Adam Petri and focuses on parallel execution strands and their dependencies. A Petri net consists of places and transitions that are connected by directed edges. Graphically, places are represented as circles and transitions as rectangles.

Every place may contain a token. Depending on the specific type of Petri net, a place can also hold more than one token. The entirety of all places and their number of tokens represents the state of the Petri net and the modeled process. The state can change by firing a transition, which consumes a token from every input place of the transition and produces a token in all of its output places.

With this basic concept, it is possible to model very complex process logic, e.g. procedures in a logistics company. Because of the simple graph structure of Petri nets, it is possible to submit a model to in-depth mathematical analysis of its properties. This designates them to be a useful tool for the professional definition of complex processes as shown by Salimifard and Wright [64]. The general applicability of Petri nets for workflow management was shown by van der Aalst [53].

Unified Modeling Language

The Unified Modeling Language (UML) was introduced in the 1990s in order to combine common modeling notations in the context of object oriented programming. The Object Management Group (OMG) started to maintain UML in 1997. Its current version is 2.4 and it has become the de facto standard for modeling of software systems [62].

UML is specified by one extensive meta model called superstructure, whose different units allow flexible modeling of both structural and behavioral patterns. Processes can be represented by different kinds of models, e.g. activity diagrams, interaction diagrams, and state diagrams [54, Chap. 2].

Interaction diagrams are used to model different participants in a process and their interaction over time. This is contrary to the necessities for modeling a pipeline that focuses on the succession of jobs and therefore this class of UML diagrams seems not ideal for the purpose at hand.

State diagrams are the graphical representation of a finite-state machine. Thus, they consist of a number of states and the transitions between them. They do not explicitly model activities, which are the most important parts in a pipeline focusing on various steps and their execution semantics.

Activity diagrams are the attempt to create a modeling technique for this purpose, with special regard to business processes. Accordingly, they support not only the explicit modeling of activities, but also data objects, different roles and other notations used in business process modeling [60, Chap. 3].

Russell et al. conclude that for most aspects of business process modeling activity diagrams are appropriate [63]. Although they name several shortcomings in modeling resources and organizational aspects, it seems that it is sufficient for the modeling of genome data processing pipelines.

Event-driven Process Chain

The Event-driven Process Chain (EPC) is a semiformal modeling notation developed 1992 by August-Wilhelm Scheer [65, Chap. 4.3]. It is an integral part of the Architecture of Integrated Information Systems (ARIS), which aims at defining a holistic modeling approach for business information systems.

The main elements of an EPC are:

- **Events**, which represent the state of a process,
- **Functions**, which represent a task or activity, and
- **Logic connectors**.

In the flow of a process, events and functions should alternate so every action results in a new state. Information objects can be used to model data flow on an abstract level and organization units can be associated to events.

EPCs are frequently used in Small and Medium-sized Enterprises (SMEs) in Germany. However, they are not adopted by an organization like the OMG.

Business Process Model and Notation

The Business Process Management Initiative (BPMI) introduced the Business Process Model and Notation (BPMN) standard in 2004. Since 2006 it is an official standard of the OMG, which released BPMN version 2.0 in 2011. Four main classes of elements exist in BPMN [65, Chap. 4.7]:

- **Flow objects** represent real world entities, such as activities, events, and gateways,
- **Artifacts** are used to model additional information, e.g. data objects, groups, or annotations,
- **Connecting objects** are any sort of line or arrow, which connect elements with each other, and
- **Swimlanes** are used to structure the responsibilities for parts of a process.

The actual work of a BPMN process is done in activity elements. They represent an atomic task that can be executed by a human or by a computer system. The logic of a BPMN flow is determined by gateways. The most basic ones are the parallel gateways representing a logical AND and exclusive gateways representing a logical

XOR. Every BPMN process needs a starting event and at least one end event to mark the start and end of the process flow [65, Chap. 4.7].

There are further concepts in BPMN. Those, which are relevant for the execution of GDP pipelines are introduced in Section 2.3.2.

BPMN is widely used for the explicit modeling of business processes and it is well suited for integration tasks performed by a computer system. A complete formal execution semantic for BPMN is still in development [67, 55]. A number of workflow management systems already offer the automated or semi-automated execution of business processes. The applicability of these systems for clinical use has already been shown [58]. The Workflow Management Coalition (WFMC) was founded in 1993 to collect and standardize workflow related activities.

A serialization for BPMN can be found in the XML Process Definition Language. XPDL is a language for the machine-readable description of processes based on XML. It is a WFMC standard and designed to support direct storage of all elements of a BPMN model.

Further Approaches

Different genome browsers offer intuitive visualization of genome data. Many offer simple analyses to be executed on the data, e.g. to find interesting elements in a set of genome data. Examples for this are the tools offered by the National Center for Biotechnology Information (NCBI) described by Wheeler et al. or the ENSEMBL project [56, 66].

However, researchers usually need more detailed analysis features for their own data. The idea of offering them a possibility for these analyses without the need of programming skills is pursued by the Galaxy project as described by Goecks et al. [57]. Galaxy is an open source project developed by the Pennsylvania State University, the Emory University, and an active community.

The concept for the use of Galaxy is based on so-called tool elements. A tool represents an individual analysis that can be executed on a set of input data. Tools can be configured and executed by the user. To carry out more complex operations, it is possible to arrange several tools in a sequence to form a workflow. Workflows are designed in a custom editor provided by the project software. They are stored in the system and can be reused in future projects.

2.3 Modeling of Genome Data Processing Pipelines

In this chapter, I describe the basic requirements for the modeling of Genome Data Processing (GDP) pipelines. The suitability of different process modeling notations is assessed and a modeling schema for these pipelines in BPMN is introduced. Furthermore, I describe how these design elements are stored in XPDL for the exchange between different parties. I selected BPMN for modeling because of its suitable collection of objects as well as its usability and intuitivity. XPDL was chosen for persistence as it is an established standard format for BPMN models.

2.3.1 Requirements Engineering

In this section, I define the requirements of a modeling notation for GDP pipelines. I describe the elements needed to model the process flow and outline the concept of pipeline models and their instances.

Modeling of Execution Information for Pipelines

The smallest unit to be referenced in a genome data processing pipeline is a job. This term refers to an actual script executed on the server to perform a specific task while activity describes the abstract representation of a job in a process model. Thus, the most fundamental precondition for modeling of GDP pipelines is a representation of a number of jobs and their execution sequence.

Often a number of jobs can be seen as a logically associated group with one discernible purpose. In most cases these composite jobs can be used in several pipelines. To enable the reuse of composite jobs and provide an additional benefit of modeling, an implementation of a modeling system for these pipelines should support the definition and inclusion of sub pipelines. In every pipeline, it should be possible to call another pipeline instead of a job. This provides users with the possibility to arrange a pipeline in a hierarchical fashion, which removes the necessity to model a sequence several times, and reduces the complexity of models.

For the efficient execution of an alignment pipeline, it is essential that parallel data processing is used whenever possible. This has to include parallel execution of different jobs as well as multiple instances of one job or sub process.

There are jobs that have a varying internal behavior or outcome depending on input parameters. For example, an alignment behavior might support a dynamic parameter for the reference genome that is to be used for comparison. The parameter would be specified during construction of a pipeline using that job. This approach could reduce redundancies in different job definitions. For this to work there must be an option to model input parameters for the jobs of a pipeline.

Model and Instance of a Pipeline

In order to exploit the whole range of possibilities that arise when using a model-based workflow system, it is useful to distinguish between the model and the instance of a process pipeline. This distinction is necessary when a model contains variable parts. Without the specification of such parts, a pipeline cannot be executed and different values for the variables can result in significantly different execution semantics. Hence a model with variables represents a group of pipelines or parts of them, comparable to a blueprint. In order to create an executable pipeline, the variables have to be defined. The model and instance of a pipeline further need to be distinguished from the process instance, which represents the actual execution of a pipeline instance. It should be possible to start with different input data in every run of a pipeline instance.

While the introduction of modeling aims to reduce the complexity and effort to create a new pipeline, the main purpose of variables is to enable the reuse of models. In the context of GDP pipelines a main process contains activities that can be done by a variety of different algorithms. Only the main pipeline has to be modeled once with a placeholder for these parts. For example, if in a pipeline various algorithms can do a specific task, only a generic term for the type of algorithm would be inserted before letting the user choose between all available algorithms of this type. A precondition for the usability of such a system is the existence of a classification of pipelines. This system has to enable the user to insert a name for a class of pipeline models that can be used for the calculation of the results for this step.

Similar to sub processes, parameters only show their full benefit, when they can be set dynamically. Some parameters can have fundamental impact on the execution semantics and on the result of a pipeline, e.g. the selected reference genome. Consequently, different values for the parameter set of a pipeline model should be accounted for as the definition of a new pipeline instance, not merely as different input parameters for the process execution.

Both sub processes and job parameters need to be available in a static and a fixed version. Thus, an implementation of such a system needs to find a way to label these spots. It also needs to provide an easy to use UI for the conversion of a set of models into a pipeline instance.

Formal Requirements

Apart from the possibility to model the concepts described in the previous sections, in my opinion, a technique for modeling of GDP pipelines has to fulfill the following criteria:

- Intuitive graphical notation,
- Prevalence of the notation, and
- Standardized machine readable representation.

The subject of the intuitive usability of a notation is heavily dependent on the context and its intended use. In this case, the criterion is that the described pipeline elements are not only supported but can be modeled directly in a simple unambiguous way that matches the intended use. For example, forks of the data flow should be obvious at a glance and parameters for jobs should be modeled in a way that directly indicates their function. This is important because it cannot be assumed that all people professionally working with genome analysis are familiar with the specifics of all modeling notations. Every element with a complex representation in the model increases the risk of errors and reduces the usability of the system.

As the aim of this work is to create a system that allows different people to execute pipeline models on a central server, it is necessary that the modeling language has got a certain degree of popularity. An important criterion for this is the existence of advanced modeling tools for the selected language. Additionally, a standardized machine-readable representation of the graphical notation is essential to assure correct interpretation of models when sharing them between different institutions.

2.3.2 Modeling of Execution Semantics

According to the requirements specified above, it is possible to assess the usability of the different modeling languages for the activities at hand. Table 2.1 summarizes the key attributes for all of the notations introduced previously.

Name	Purpose	Std	GDPP
Petri Nets	Mathematically verifiable sequence modeling	✗	✗
EPC	Business processes	✗	✗
UML	Technical specification of software characteristics	✓	✓
BPMN	Detailed modeling of business processes and workflows	✓	✓

Table 2.1: Summary of evaluated modeling notations and their applicability for GDP pipeline modeling (Std = Standardized, GDPP = GDP pipelines, ✗= No or insufficient support, ✓= Full support).

With the definition of flow and data elements that I presume necessary for the modeling of GDP pipelines, Petri nets and EPCs can be ruled out as conclusive modeling techniques. Both have insufficient support for parallel execution of activities with a variable number of instances. Petri nets also do not provide methods for the modeling of data objects and sub processes. Additionally, both notations lack a standardized machine-readable representation that would enable sharing of models between different institutions and EPCs are not widely used except in Germany.

In contrast to the previous examples, BPMN and UML activity diagrams support all the elements described previously. They both have at least one standardized XML representation. The graphical and logical elements are very similar to each other. A distinction can be made by the general orientation of the techniques. While activity diagrams aim for the extensive specification of software flow, BPMN focuses on the modeling of automated and manual business processes and intuitive usability by non-experts.

The question, which of these techniques is more fitting for different use cases is subject to numerous scientific papers [59]. The decision depends strongly on the specific situation. For the activity at hand, the main focus should lie on easy modeling of key elements. These are parallel activities or sub processes and input parameters for jobs. For the latter, there is a difference in the way of modeling. In activity diagrams, data objects are part of the process flow and usually simultaneously output of one activity and input for another. In contrast, BPMN shows input and output data as data objects associated with one or several activities independent from the process flow. This offers the possibility to store input parameters as data objects that are input parameters of an activity without being the output of a previous activity.

With these differences in mind, BPMN has two advantages: a more fitting concept for the representation of data objects for this case and the benefit of being developed with special focus on usability and intuitive modeling elements. For this reason, I chose BPMN as modeling language in the remainder of my work although UML activity diagrams might also be a promising candidate.

Process Flow

Every job to be executed in the pipeline is represented by an activity element in BPMN. Although a job might have several predecessors or successors, every activity should only have one incoming and one outgoing flow in BPMN. Every process pipeline should have one starting and one end event. Figure 2.1 shows a basic form of a pipeline.

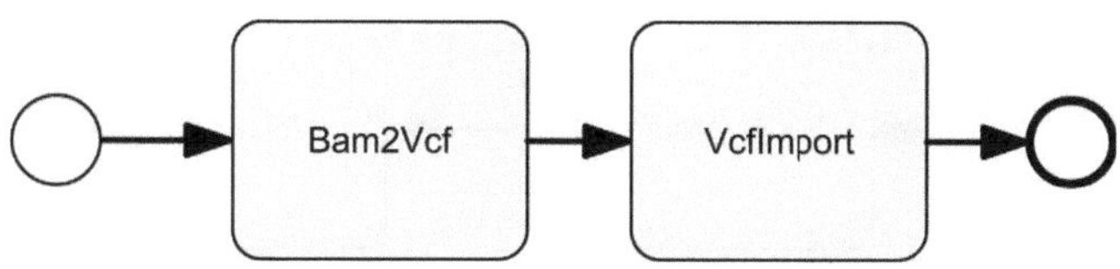

Fig. 2.1: Basic principle of a genome data processing pipeline modeled in BPMN.

Hierarchy of Jobs

The processing pipelines can be nested hierarchically to any depth desired. Any set of logically associated jobs can be represented as a separate file containing a process diagram. In the invoking process a sub process activity is used as a placeholder for the nested jobs. The name of that activity has to be the same as the name of the sub process in order to automatically insert it in the parsing process. Figure 2.2 shows the usual depiction of a sub process in BPMN.

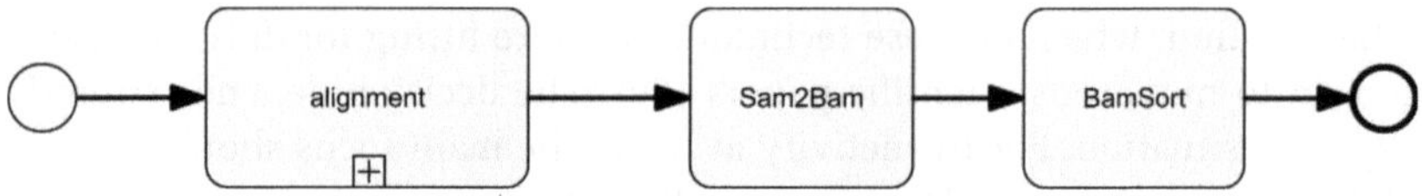

Fig. 2.2: Modeling of a sub process.

Parallel Execution of Jobs

There are two ways of defining the parallel execution of one or several jobs. The first one is using parallel gateways. When a parallel gateway is signaled, all outgoing edges of the gateway are signaled. When used with one incoming and several outgoing edges, the gateway only signals after all incoming edges are activated. So the sequence flow can be split in two or more parallel strands and re-synchronized if needed. An example for this is shown in Figure 2.3.

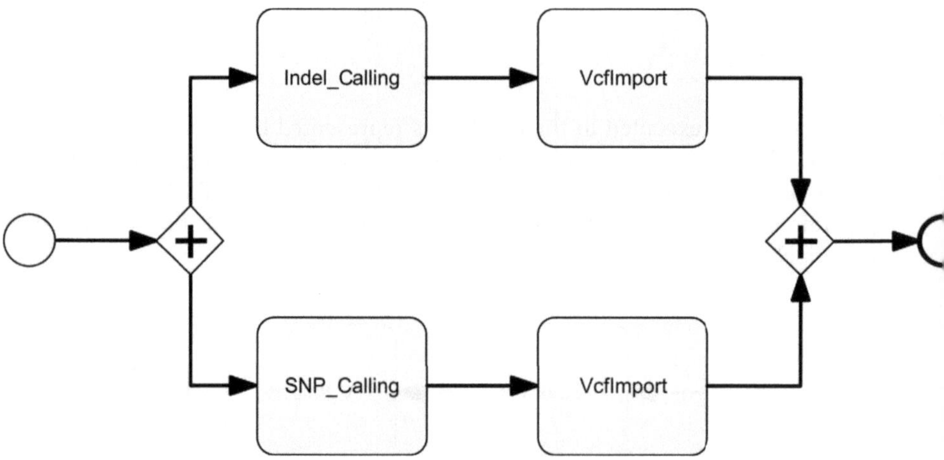

Fig. 2.3: Modeling of parallel gateways.

Often it is necessary to execute one job simultaneously several times. In that case it is preferable to use a parallel multiple instance activity, which is depicted as an activity with three vertical lines at the bottom as shown in Figure 2.4. In the BPMN standard, there are several ways to define the number of instances of a multiple instance activity [61, Sect. 13.2].

These possibilities are implemented in XPDL differently, depending on the modeling tool used. Additionally, most tools only allow the insertion of numeric values in the according fields. This is problematic when modeling a sub process with a variable number of instances as described in Section 2.3.2.

In my system the number of job instances to be started is included in square brackets behind the name of the activity. This circumvents problems arising from

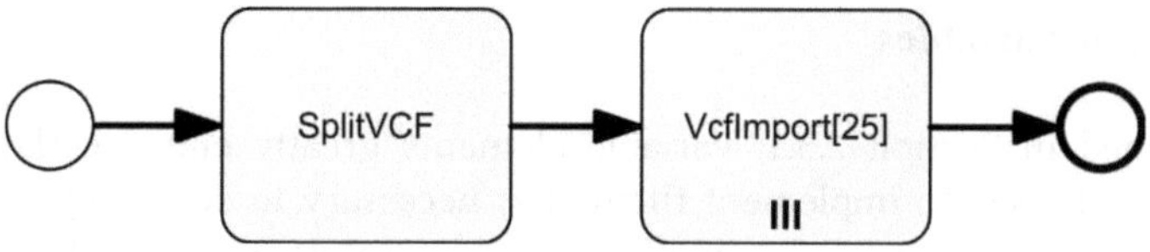

Fig. 2.4: Modeling of a multiple instance parallel activity.

different implementations of modeling programs and users are spared the configuration of the complex input parameters.

A multiple instance activity can also be a sub process. Thus, it is possible to execute complex control flows simultaneously. Those two ways can also be combined, for example if two different jobs have to be executed several times and they can run in parallel.

Modeling of Parameters

BPMN provides a very fitting way to model input parameters for jobs by using data objects as input for activities. A parameter is stored in a data object with the parameter name in front, followed by a colon and the value of the parameter. The parameter name matches the input variable name of the corresponding job. A parameter in form of a data object can be associated to one or several activities. Figure 2.5 shows how the variable `reference` with the value `hg19` is defined.

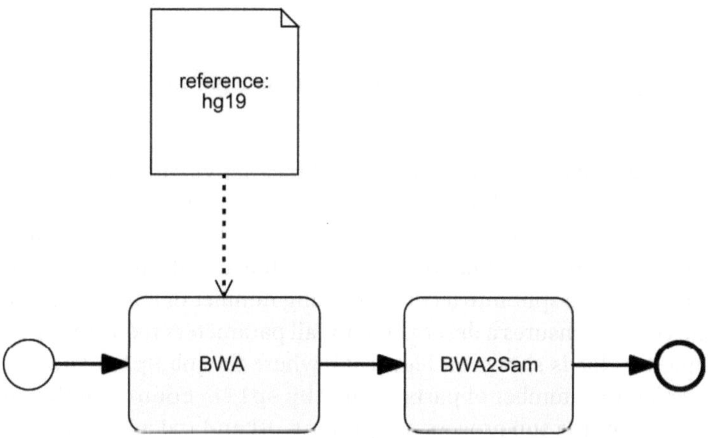

Fig. 2.5: Modeling of parameters as input for an activity.

Modeling of Variables

As described in Section 2.3.1, variable elements greatly enhance the profit of pipeline modeling. To implement them, it is necessary to define an easy way of marking all variable parts of a pipeline. For this prototype a variable is indicated by a $ preceding the variable's name. This can be used for job parameters, sub processes and the number of instances of a multiple instance activity alike.

For variable job parameters it is possible to define a description of the parameter that is shown when the pipeline is configured. A colon indicates the boundary between the description and the variable name the same way, the value would be defined for a fixed parameter. In Figure 2.6 the previous example is shown as a variable parameter with the description `reference genome` instead of a value.

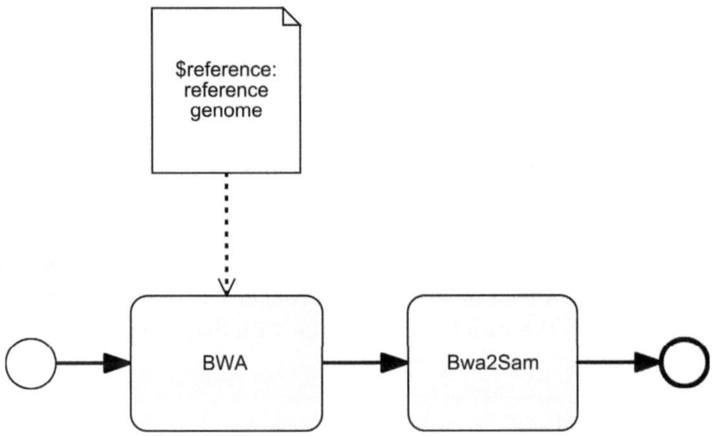

Fig. 2.6: Declaration of a variable parameter.

The number of instances of a multiple instance activity might sometimes not be fixed when modeling a pipeline. In these cases, it is possible to insert variable name lead by a $ in place of a number. This variable has to be referenced as a parameter somewhere in the pipeline. In most cases, the activity preceding the current one where the data is split into a corresponding number of parts is chosen for this. Hence, the concept ensures a description for all parameters requiring specification. An example for this is shown in Figure 2.7 where the job `SplitFastQ` splits the working data into a number of parts defined by `split_count` and the same number of instances of the sub process `Alignment_Standard` are started afterwards.

A sub process is marked as variable by a $ preceding the name. The name following the $ is interpreted as the description for a type of sub processes as contained in the corresponding database table `Models`. In the example, the variant calling can be done by different means. The selection of one specific process is part of the pipeline instance definition.

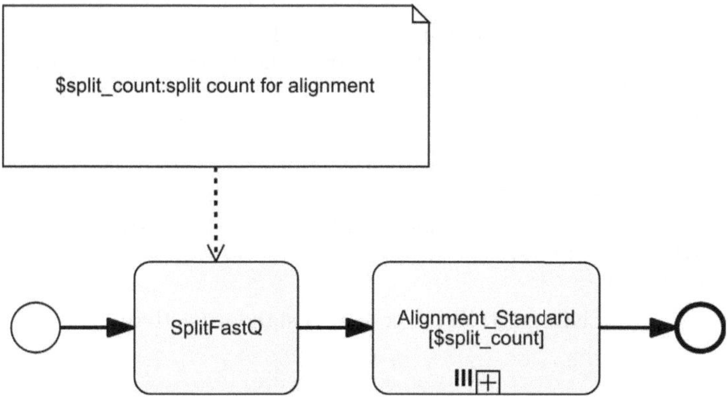

Fig. 2.7: Modeling of a sub process with a variable number of instances stored in split_count .

If a parameter is supposed to be configurable, the name is lead by a $ accordingly. This signals the parser that the text behind the colon is interpreted as a description of the variable, which can be used to ask users for a specific value.

2.3.3 *Machine Readable Model Representation*

In order to automatically interpret a BPMN model, it is essential that there is a standardized machine-readable representation of the graphical notation. A very widespread and well-defined XML representation of BPMN models is XPDL. In this section, I outline how XPDL can store the BPMN elements as described above.

Representation of Pipeline Elements in XPDL

The most important XPDL tag is `Activity` tag. It offers various opportunities for specification and represents most elements of a BPMN model. In its most basic form it represents a standard activity as shown in Listing 2.1.

A sub process differs from a standard activity only by the `Implementation` tag. It contains a `SubFlow` tag where a reference to the sub process and various other specifications can be defined as shown in Listing 2.2. Because the form of link between models vary between different modeling tools and are often only usable inside these tools, a link specified at this point is not taken into account for this function. A more important factor is the field `View` that determines whether a sub process symbol is shown for this activity or the content of the sub process is defined in the same model inside an extended activity shape. As only the first

```
<xpdl:Activity CompletionQuantity="1" Id="newpkg1_wp1_act2"
    IsATransaction="false" IsForCompensation="false" Name="BWA"
    StartQuantity="1" Status="None">
    <xpdl:Implementation>
        <xpdl:No/>
    </xpdl:Implementation>
    <xpdl:Performers>
        <xpdl:Performer>newpkg1_wp1_par1</xpdl:Performer>
    </xpdl:Performers>
</xpdl:Activity>
```

Listing 2.1: XPDL code of a standard activity

```
<xpdl:Implementation>
    <xpdl:SubFlow Id="" View="COLLAPSED">
        <xpdl:EndPoint EndPointType="WSDL">
            <xpdl:ExternalReference location="http://"/>
        </xpdl:EndPoint>
    </xpdl:SubFlow>
</xpdl:Implementation>
```

Listing 2.2: XPDL code for a sub process

of these options is supported in the research prototype I present, this parameter should always be set to `collapsed`.

With a `Loop` tag inside the activity specification, a multiple instance activity can be defined. The number of instances to be started and the exact terms for their execution can be defined in this tag and the superior `Activity` element. This reflects the complex adjustments specified in BPMN. But for the execution of genome data processing pipelines it usually is sufficient only to specify the number of instances of an activity without any extra information. Additionally, when comparing different tools, the XPDL standard is sometimes implemented in different ways so several possible forms of storing the value for the number of instances would have to be considered when interpreting an XPDL file.

So the number of instances is stored in the activity name as described above. The important thing about the `Loop` tag is the correct setting of the parameter `LoopType` to `MultiInstance`.

Gateways are also represented by an `Activity` tag in XPDL. They do not possess a name attribute, but are very similar to standard activities. They differ by the tags `Route` and `TransitionRestriction` that are used to define the exact behavior of the gateway. A parallel gateway is sufficient for all pipelines examined in the course of the project and therefore the only form of gateways supported by the prototype. For this very simple gateway, no special adjustments have to be taken into account. So the only parameter that has to be processed is `GatewayType`, which has to be set to `Parallel` as shown in the first line of Listing 2.3

```
<xpdl:Route ExclusiveType="Data" GatewayType="Parallel"
    Instantiate="false" MarkerVisible="false"/>
<xpdl:TransitionRestrictions>
    <xpdl:TransitionRestriction>
        <xpdl:Join ExclusiveType="Data"/>
        <xpdl:Split ExclusiveType="Data" Type="Inclusive">
            <xpdl:TransitionRefs>
                <xpdl:TransitionRef Id="newpkg1_wp1_tra2"/>
                <xpdl:TransitionRef Id="newpkg1_wp1_tra3"/>
            </xpdl:TransitionRefs>
        </xpdl:Split>
    </xpdl:TransitionRestriction>
</xpdl:TransitionRestrictions>
```

Listing 2.3: XPDL code for a parallel gateway

```
<xpdl:Transition From="newpkg1_wp1_act5" Id="newpkg1_wp1_tra6" To
    ="newpkg1_wp1_act6">
</xpdl:Transition>
```

Listing 2.4: XPDL code for a transition between two activity nodes

```
<xpdl:Artifact ArtifactType="DataObject" Id="newpkg1_1" Name="
    newpkg1_1">
    <xpdl:DataObject Id="newpkg1_1" Name="reference:hg19"/>
</xpdl:Artifact>
```

Listing 2.5: XPDL code of a data object

```
<xpdl:Association AssociationDirection="From" Id="newpkg1_1"
    Source="newpkg1_1" Target="newpkg1_wp1_act2">
</xpdl:Association>
```

Listing 2.6: XPDL code for the association of a data object to an activity

A transition between two elements of the process flow references them by the `Id` attribute in every `Activity` tag.

Data objects are a special form of the class of artifacts that also includes, for example, comments. They are identified by the parameter `ArtifactType` that has to be set to `DataObject`. The content of the data object is stored in a separate sub tag as shown in Listing 2.5. An association from a data object to an activity is established similarly to a transition as shown in Listing 2.6.

All described elements contain additional graphics information to specify the way they are displayed. They may also contain several other configuration options. With the basic set of nodes shown here, the execution semantics of genome data processing pipelines can be fully expressed.

2.4 Application Example

As part of the underlying project, I designed a prototype for the interpretation of pipeline models specified in XPDL according to the schema described in Section 2.3.2. In this chapter, I outline its functionality and internal structure.

There are two target groups for the system. The first are the system administrators, who set up the jobs and pipelines accessible for external users. They may still alter or add jobs. For them the only change is that the structure of pipelines is no longer hard coded in the job scripts. For all tasks concerning pipeline composition, they can use the same procedure as the second group, which is composed of external researchers and bioinformaticians. The abstraction added by the modeling allows anyone to design and execute pipelines with the jobs available on the server without accessing the source code.

In contrast to these two groups, a physician or medical expert does not have the knowledge to design a customized pipeline. They use verified standard pipelines offered by the system. Still, they can profit from the new approach. Although they cannot design models for themselves, they may configure individual pipeline instances. This is done via the project's web service. If the models offered there are designed correctly, a user can fill out the variable parts without knowledge of the internal working of a pipeline.

2.4.1 Pipeline Configuration User Interface

The user interface for pipeline configuration is part of a web service developed for the High-performance In-memory Genome (HIG) project described in Section 1.5.

The main challenge for the design of the user interface was the desired usability for both researchers configuring complex custom pipelines, and physicians or medical staff. The latter should be able to assemble a pipeline instance from given standard pipelines without the risk of a misconfiguration invalidating the results.

To achieve this, a minimum number of elements is shown at each instant. In the beginning, only a field for input of a pipeline name and a drop-down menu for the selection of the main pipeline are shown. When a main pipeline is selected, the web service executes a lightweight parser that returns the structure of the pipeline as it is defined at the current moment and a list of all variable parts of that pipeline. Those variables are then presented to the user for specification as shown in Figure 2.8. Parameters are given as text input while each variable sub process is shown as a drop-down menu that contains all models fitting in that place, according to the type specified in the model.

During configuration of the pipeline instance, its current state is always displayed at the bottom of the UI. Thus, a user gets direct feedback about what he is doing. This visualization shows the process flow and the interleaving structure of sub processes as depicted in Figure 2.9. If an activity or sub process is executed several times, the number of instances is displayed at the bottom, as is done if the

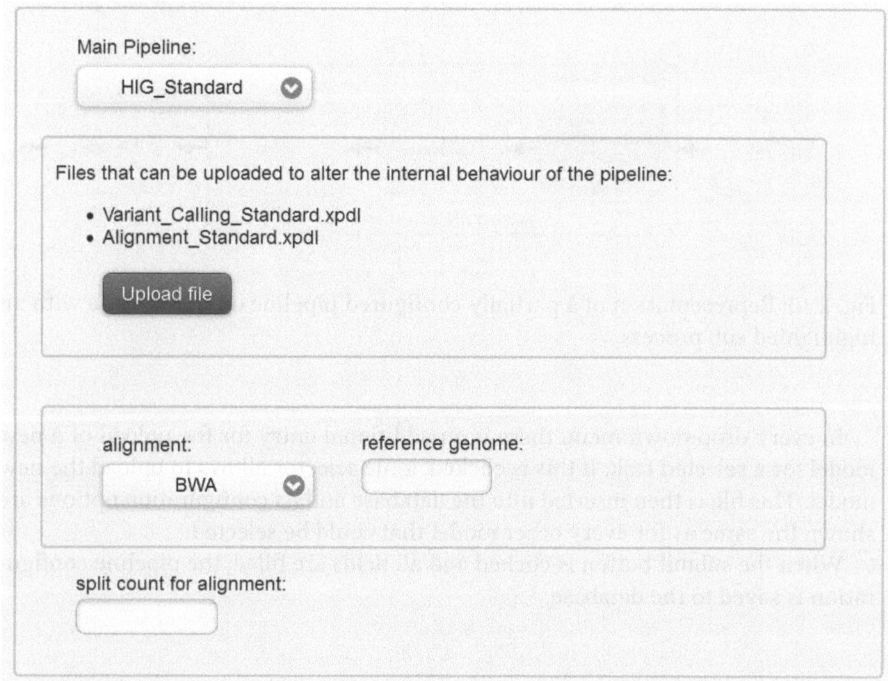

Fig. 2.8: Screenshot of the pipeline configuration front end.

number of instances is variable. In that case the corresponding variable name is shown.

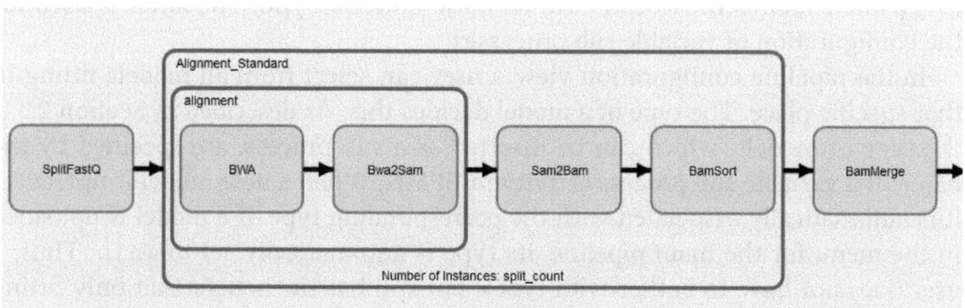

Fig. 2.9: Representation of a fully configured pipeline on the website with the sub processes indicated by grey rectangles.

When the user does not yet define a variable sub process, the corresponding activity in the visualization is colored red with a note, indicating it has to be specified before the pipeline instance can be created. This is shown in Figure 2.10.

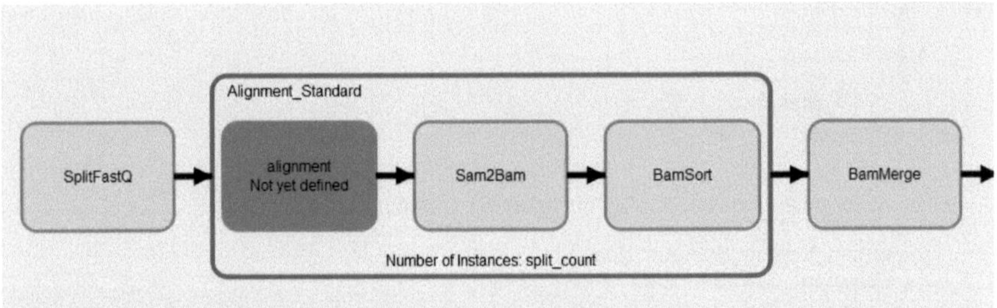

Fig. 2.10: Representation of a partially configured pipeline on the website with an highlighted sub process.

In every drop-down menu there is an additional entry for the upload of a new model for a selected task. If this is clicked, a file selector allows to upload the new model. This file is then inserted into the database and its configuration options are shown the same as for every other model that could be selected.

When the submit button is clicked and all fields are filled, the pipeline configuration is saved to the database.

2.4.2 Data Format for Pipelines

For the administration of the pipelines, I created two tables in the database: `Models` and `Pipelines`. The `Models` database schema contains the XPDL representation of the pipeline models. They are stored in a column with Character Large Object (CLOB) data type that is used for the storage of big character data. Above this, every line contains the `model id`, its `name` and the `type`. The latter is used for the configuration of variable sub processes.

In the pipeline configuration view a user can select from all models fitting in that specific place. The type of a model decides this. As described in Section 2.3.2, the type of models, which can be inserted as a sub process, are specified by the name of a variable sub process activity in BPMN. When a new model is uploaded, it is automatically associated with the corresponding type. If a model is uploaded in the menu for the main pipeline, its type is automatically set to `main`. Thus, a user does not have to bother with types, but still has the benefit that only fitting models are offered to him.

In the `Pipelines` schema, the actual models used in a pipeline instance are referenced only by their identifier. The main model is stored as an integer data type while all subordinate models are stored as a JavaScript Object Notation (JSON) string, representing an associative array. This maps each sub process existing in the pipeline instance to the `id` of the model that was chosen for this task. Similarly, the parameter configuration is stored as an associative array mapping each variable name specified in the BPMN data objects to its value. Additional columns are used for the `id` and the `name` of a pipeline instance.

For the actual execution of a pipeline instance, the models mentioned in the pipeline configuration have to be assembled and brought into a form that directly specifies what is the actual succession of jobs. The parser class written in Python implements this.

It returns a pipeline object as the representation of a process instance of a pipeline. That consists of several step objects that each represents one instance of a job to be executed in the run of this pipeline. Every instance of the step class has a set of previous steps that have to be finished for its start and a set of next steps that wait for its completion. Additionally, a step can have any number of parameters that are given to the job upon starting time.

A pipeline can also have several strands. Each strand instance represents one straight part of a pipeline that does not include any forks. For the parsing of a pipeline instance strands are used to support the handling of pipelines.

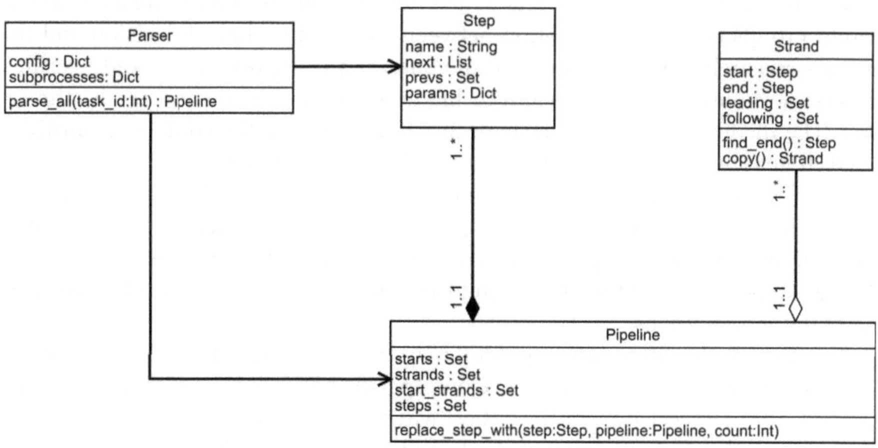

Fig. 2.11: UML class diagram representing the architecture used for parsing a pipeline instance. Only a subset of attributes and methods are shown.

The parser class creates a step object for every activity in a model. In the case of multiple instance activities a step object is created for every instance. When a model contains a sub process, the parsing is done recursively to include hierarchically nested pipeline jobs. An instance of the parser class is responsible for the processing of one model and returns a pipeline object that contains the representation of this model and of all sub processes. The pipeline object returned by the sub parser is then inserted in place of the original step representing the sub process activity using the pipeline class method `replace_step_with`.

The complete pipeline object is handed over to the worker framework responsible for the execution of the jobs. This framework and the scheduling of jobs are described in detail in Chapter 3.

2.5 Evaluation and Discussion

To assess the benefit, which can be gained by the proposed system, I conducted user interviews with bioinformaticians. In this section, I will consider different aspects of a system offering genome analysis that were mentioned as crucial, and analyze my contribution with regards to them.

Distinction Between Clinical Use and Research

One of the most important aspects of a system that offers genome analysis to different groups is the differentiation between the use for clinical decision and the use for research purposes. In a clinical context, it is important that only verified pipelines are used since the health or life of a patient may be affected by the result. The medical staff cannot be expected to understand the complex algorithms used in genome data processing pipelines. This usage is supported, for example, by the HIG platform, where users can choose between several fixed pipelines for the analysis of a file. A model-based approach can be enhanced by the possibility to configure individual pipeline instances using verified models from the server. These can be fitted together and configured without the risk of disrupting the correct interaction of jobs.

In contrast, algorithms and pipelines may change several times a year and often it is important for a researcher to use one specific pipeline for a task in a research context. For most researchers, the easiest way to operate is to work on a self-owned computer with self-written scripts that can be adapted at will.

Full Flexibility

To get closer to the goal of an integrated platform that brings together researchers, physicians and patients, it is essential to provide extensive control for researchers over the configuration of their pipelines. Even if a central server offers more technical resources and thus a faster processing of tasks, it can be assumed that many researchers could not use it for their main purpose without this flexibility.

The modeling of pipelines according to my proposition allows this flexibility since not only the succession of jobs can be specified but also any number of parameters for detailed configuration of jobs. Its applicability depends on two conditions: The regular update of the set of jobs available on the system and the provision of an extensive documentation of their interfaces and configuration options.

If these requirements are met, it is possible to enhance the value of a system like the HIG platform considerably for researchers and thus support the vision of bringing together all groups concerned with the treatment of genome related diseases. This can be seen as the major accomplishment of a model-based approach for pipeline definition.

Detailed Logging

For researchers it is essential to guarantee that all information about the processing of genome data is retrievable afterwards because this information has to be included in publications to ensure conformability of scientific findings. This aspect was out of scope of this contribution. However, the prototype lays a foundation for detailed logging as all pipeline instances used are stored in the database. By using the features for analyzing historical data provided by the in-memory database, this approach could be developed further. Additionally, all job scripts have to implement features for this. The logs should be designed to be compatible with current standards for laboratory documentation, in order to provide detailed logging of individual steps.

Partial Pipeline Execution

Often it is necessary to execute parts of a pipeline, e.g. in order to retry an algorithm with different parameters on data, which was already preprocessed with a notable effort. In the prototype, a second pipeline with parts of the steps missing would have to be designed in these cases. On one hand, this is additional work for the user but on the other hand, this supports the traceability of actions on the system.

Comparison to the Galaxy Project

The main difference between the approach introduced here and the Galaxy project, is the utilization of a standardized modeling notation. Where Galaxy offers a customized tool for the modeling of workflows, I propose to make use of BPMN as an open notation that is subject of a lot of research. One major benefit of this is the facilitation of sharing pipelines, not only within one platform but also between different platforms and institutions. BPMN also offers a well-defined and intuitive variety of flow elements, exceeding those used until now in genome analysis pipelines. Hereby it offers the opportunity for further development of pipelines to make the process of genome analysis more efficient or easier to handle.

The modeling approach with activities that directly map to a script on the server is closer to the technical background since the underlying script could make all parameters re-definable, with the user picking those he wants to alter from their standard values. Although this concept of overwriting parameters is not implemented in the prototype, it is easily compatible with the system. It could make the system more attractive, not only for biological researchers without programming knowledge, but also for bioinformaticians and technicians, who could get nearly as much control over the script execution as they would have on their own machine.

On the other hand, medical staff could be reached as a second new target group that usually does not profit from Galaxy. They could have the benefit of configuring pipelines without the risk of invalidating a pipeline and without the need for in-depth knowledge of the inner working of the pipeline.

2.6 Conclusion and Outlook

A model-based approach to the offering of genome data processing pipelines can greatly enhance the value of a platform for different target groups compared to the use of static pipelines. Especially if it is the aim to bring together different groups such as researchers, physicists and patients on one platform, it is important to offer an easy to use, multi-level way to define analysis pipelines.

The further development of this approach could lead to a system, which is oriented on classical workflow engines. Pipelines could provide choices between two different paths to take. These choices could be determined by simple conditions on result values or even by the user. For this case, BPMN offers the possibility to define user activities. These could be integrated into pipelines, in places where a decision has to be taken based on the data that only a researcher with background knowledge can evaluate. The user could then be alerted whenever a decision for a pipeline has to be taken. This would make situations much easier where parts of a pipeline have to be recalculated under certain circumstances or a different algorithm is chosen for further processing according to previous results. In such a situation otherwise several different pipelines would have to be designed and started manually.

2.7 References

[53] van der Aalst WMP (1998) The Application of Petri Nets to Workflow Management. The Journal of Circuits, Systems and Computers 8(1):21–66

[54] Booch G, Rumbaugh J, Jacobson I (1998) The Unified Modeling Language User Guide, 1st edn. Addison-Wesley, Reading, MA

[55] Dijkman RM, Gorp PV (2011) BPMN 2.0 Execution Semantics Formalized as Graph Rewrite Rules. In: Business Process Modeling Notation, Lecture Notes in Business Information Processing, vol 67, Springer, pp 16–30

[56] Flicek P et al. (2012) Ensembl 2013. Nucleic Acids Research 41(D1):D48–D55

[57] Goecks J, Nekrutenko A, Taylor J (2010) Galaxy: A Comprehensive Approach for Supporting Accessible, Reproducible, and Transparent Computational Research in the Life Sciences. Genome Biology 11(8)

[58] Huser V et al. (2011) Implementation of Workflow Engine Technology to Deliver Basic Clinical Decision Support Functionality. BMC Medical Research Methodology 11(1):43

[59] Ko RKL, Lee SSG, Lee EW (2009) Business Process Management (BPM) Standards: A Survey. Business Process Management Journal 15(5):744–791

[60] Marshall C (1999) Enterprise Modeling with UML - Designing Successful Software Through Business Analysis. Addison-Wesley

[61] Object Management Group (2011) Business Process Model and Notation (BPMN). http://www.omg.org/spec/BPMN/2.0/PDF/. Accessed Sep 23, 2013

[62] Rumbaugh J, Jacobson I, Booch G (2004) Unified Modeling Language Reference Manual, The (2nd Edition). Pearson Higher Education

[63] Russell N et al. (2006) On the Suitability of UML 2.0 Activity Diagrams for Business Process Modelling. In: Proceedings of the 3rd Asia-Pacific Conference on Conceptual Modelling, Australian Computer Society, Hobart, Australia, vol 53, pp 95–104

[64] Salimifard K, Wright M (2001) Petri Net-based Modelling of Workflow Systems: An Overview. European Journal of Operational Research 134(3):664–676

[65] Weske M (2007) Business Process Management - Concepts, Languages, Architectures. Springer

[66] Wheeler DL et al. (2008) Database Resources of the National Center for Biotechnology Information. Nucleic Acids Research 36:D13–D21

[67] Wong PYH, Gibbons J (2008) A Process Semantics for BPMN. In: Liu S, Maibaum T, Araki K (eds) Formal Methods and Software Engineering, no. 5256 in Lecture Notes in Computer Science, Springer Berlin Heidelberg, pp 355–374

Chapter 3
Scheduling and Execution of Genome Data Processing Pipelines

Cornelius Bock

Abstract Genome data processing pipelines extract a list of mutated genes out of raw genome data, which is an essential information required by physicians and researchers from a patient's genome. Improving performance, stability and flexibility of pipeline execution environments is therefore a matter of importance. Today, analysing a sequenced human genome takes up to several weeks. In scientific environments, parallelization using computer clusters and in-memory technology can accelerate execution to several hours, but flexibility and scheduling need to be improved, too. In this work, I propose a way to execute diverse genome data processing pipelines on a cluster of worker machines coordinated by a single scheduler using an in-memory database. I will show how a scheduling algorithm, which estimates processing time by analyzing execution logs, improves the systems throughput. I will also discuss how the database can be used as communication medium, log, decision instance and statistics service and how the system can benefit from its power.

3.1 Introduction

A list of variants in genes is vital information for physicians and researchers as described in Section 1.1. Extracting this list out of raw genome data is the task of Genome Data Processing (GDP) pipelines.

There are two major jobs to be done to achieve this: alignment and variant calling, as described in Section 1.1. Both work not just with the data of the genome, which is processed, but also include huge datasets of information around it, like reference genomes or datasets about investigated mutations. Around those essential jobs there are several other steps needed for execution, e.g. pre- and post-processing, format conversions, sorting of results, and quality assurance. They differ in the algorithms they use, application of auxiliary tasks, and the version of the reference genome. GDP pipelines process FASTQ files, which contain snippets of sequenced base pairs of the human genome as described in further detail in Chapter 2 [71].

Next-generation sequencing accelerated the sequencing of genomes significantly [69]. Thus, GDP pipelines become more and more relevant for physicians to treat patients based on information from their genome. Researchers try to improve the quality of results by validating them multiple times and adding more quality assurance steps, which makes GDP pipelines more and more complex [77]. Improving performance, stability and flexibility of pipeline execution environments remains therefore a matter of importance.

In this work, I propose how an In-Memory Database (IMDB) distributed on a cluster of worker nodes achieves these goals. Jobs will be apportioned between the nodes to exploit their computational and memory capabilities using parallelization. I constitute a single organization entity that is responsible for the coordination of pipeline executions and schedule them to achieve maximal throughput. Therefore, I discuss the use of a scheduling algorithm, which estimates remaining execution time using execution logs in order to process shortest pipelines first.

Always executing the shortest remaining task first will lead to optimal throughput and turnaround time, which is discussed in more detail in Section 3.3.3. However, determining the shortest task among a set of possible tasks is not trivial. For a precise estimation, one can simulate execution to measure execution time. Nevertheless, this would increase scheduling overhead to be greater than its benefit. Therefore, I propose a scheduling algorithm, which uses the capabilities of an IMDB for estimating execution time. I claim that this scheduling algorithm takes less than one second to reach a scheduling decision while analyzing a history of 2,000 pipeline executions containing about 300,000 database rows. I also claim that this speed grows linearly with the size of the execution history.

The rest of this work is structured as follows: In Section 3.2, I will demonstrate scheduling algorithms used in operating systems. Adaption of these scheduling algorithms and their usage of in-memory key-concepts are discussed in Section 3.3. I describe in Section 3.4, how they are used in the research prototype and how pipeline flexibility is established. My findings are discussed in Section 3.5 and my work concludes in Section 3.6.

3.2 Related Work

In the following, I will describe the relevance of scheduling in operating systems as related work for this contribution. Tanenbaum defines the part of an operating system that makes the choice out of a set of ready processes competing for the CPU, which to run next, as the scheduler, and the algorithm it uses as the scheduling algorithm [81, Sect. 2.4]. He extensively discusses, which scheduling algorithms are used for different optimization goals and how they try to achieve this in the context of operating systems. Arranging subtasks of GDP pipelines to worker processes distributed on different nodes is analogical to arranging threads of processes to processors, so scheduling concepts of operating systems are applicable for my GDP pipeline scheduler. Hence, I will summarize Tanenbaum's findings.

He firstly differentiates schedulers by the time they work. Non-preemptive scheduling algorithms wait for a running job until it blocks and then decide for the next one. "In contrast, a preemptive scheduling algorithm picks a process and lets it run for a maximum of the fixed time [81, Sect. 2.4]." At the end of such a fixed time, a process is suspended until the scheduling algorithm chooses to run it again. This requires the possibility to interrupt a running job, save its state, and later restart it from the same point again.

Beside general objectives like fairness, policy enforcement, and balance, Tanenbaum categorizes the following environments in order to distinguish appropriate goals that scheduling algorithms are trying to accomplish:

- **Batch systems**: Throughput, turnaround time, CPU utilization,
- **Interactive systems**: Response time, proportionality, and
- **Real-time systems**: Meet deadlines, predictability.

He explains the goals as follows: Fairness warrants comparable service to comparable processes, while policy enforcement amounts implementing the system's policy. Balancing means utilizing all parts of the computer when possible, especially by alternate CPU and memory intense processes. Throughput is measured by the amount of processes finished in a specific time, whereas the average sum of waiting and executing a process is called turnaround time. Response time is very similar to the latter, but prioritizes foreground jobs in order to enhance user experience. Meeting the temporal expectations a user has from certain processes is called proportionality, e.g. when reading a file stream, which is a video from the hard disk, each frame should be loaded when it is needed for displaying, so execution of this job should be both: finished within the deadline and predictable in order to encourage regularity.

Tanenbaum defines real-time scheduling as following: "The scheduling of multiple competing processes, some or all of which have deadlines that must be met is called real-time scheduling [81, Sect. 7.5]." This is stated in the context of multimedia systems and their requirements to supply results timely, which does not refer to the time the scheduler needs but to the goals it has to achieve. This does not correspond with my definition of real-time scheduling stated in Section 3.3.3.

Ehses et al. propose two-stage scheduling [73, Sect. 9.3]. This considers swapping time in operating systems when interchanging running processes. Two-stage scheduling will not be considered within this work, since swapping does not apply as described in Section 3.3.3.

First Come First Served

With the First Come First Served (FCFS) scheduling policy, the first process ready to be executed will immediately be started and runs without interruption until it is finished or it blocks, e.g. when waiting for user input. Every time a process becomes ready, it is appended to the queue of waiting processes. When a resource becomes available, the first job from the queue is scheduled for execution.

As Tanenbaum notes, resources are not distributed well when there are multiple short running and at least one long running job. When these jobs occur in an adverse order, the long running process will block the others deteriorating the overall turnaround time.

Shortest Task First

In order to minimize turnaround time and maximize throughput, executing the Shortest Task First produces best results in most cases [81, Sect. 2.4.2]. Processes with a, b, c, and d sec pure execution time, respectively, are executed in that order. The mean turnaround time is $\frac{1}{4}(4a + 3b + 2c + d)$, because the first will finish after a sec, the second after $a + b$ sec and so on. This means that the earlier a job is executed, the bigger is its influence on the mean turnaround time, e.g. if a will take one sec more, the mean turnaround time will grow by a whole second, while the same delay in d only affects the result with a fourth of its delay.

As Tanenbaum states, this only guarantees optimal behavior when all processes start at the same time. When starting unfavorably, even FCFS results in a smaller turnaround time. This is only a problem when using non-preemptive scheduling, as the problem exacerbates because of the missing possibility to interrupt a long running task in order to give a meanwhile arrived, short task preference.

Yet, minimizing mean turnaround time is still the strength for a preemptive version of this algorithm. They always execute the process with the shortest remaining time first. Preemption offers the opportunity to suspend long remaining tasks in order to complete short ones first.

However, both versions have a great disadvantage: They need to know the execution time for any job waiting in the queue. It is not a viable option to estimate their execution time through simulation as it takes too long to offer a real scheduling benefit.

Tanenbaum proposes a way to estimate execution time, which is assuming it as the previous sum of last executions. He proposes $a = \frac{1}{2}$ as weight factor, which leads to simple aggregations. Weighting the sum makes older values more and more negligible, so history does not affect the current results.

When scheduling a distributed system, another challenge appears: Additionally to the next job to be executed, the node or processor has to be chosen. Tanenbaum addresses this problem with several work-balancing algorithms [81, Sect. 8.1.4]. He proposes a single, system-wide queue for ready tasks as the simplest solution. Every time a processor becomes available, it gets the next thread.

Some more complex algorithms, such as gang scheduling, consider the communication between threads and try to schedule them simultaneously on different processors in order to shorten waiting time for responses. Due to the fact that the entities I schedule within this work do not communicate with each other during their runtime, this methodology does not apply to GDP pipeline execution as explained in Section 3.3.

3.3 Method

A basic idea to speed up pipeline execution is to parallelize as much as possible. However, the more executing entities are involved, the more complex distributed coordination becomes. Therefore, I use a single node as coordinating instance. Together with a distributed and a set of other nodes running multiple workers each, it forms the worker framework.

The scheduler node is responsible for managing all aspects from parsing the pipeline model to scheduling all accruing jobs. How the decision is made, which job should be executed next, and how this scheduling algorithm is integrated between traditional ones is particularized, too.

I define task to be a pipeline with definite input values to be executed, and a single item of that task is a subtask. The algorithm used to execute a subtask is called job. The scheduler will use steps as its internal representation of subtasks to organize them, which is described in Section 3.4.

In Section 3.3.1, I will describe the requirements that formed the basis for architectural design and implementation decisions. Afterwards, key database features are expounded in Section 3.3.2.

3.3.1 Requirements of the Execution Environment

In order to achieve maximum flexibility for pipeline design, pipelines are provided as linked lists of step objects consolidated in a pipeline object. A parser as described in Chapter 2 provides them. I will first specify data structures and then illustrate the requirements the scheduler needs to meet when using the parser.

GDP pipelines consist of subtasks to be executed. Each of them can have input and output values, whereas the output of one subtask will be the input of its subsequent subtasks. In order to exploit parallelization benefits, there can be multiple execution strands, which have to be split and merged as modeled as UML object diagram in Figure 3.1 [68].

The scheduler will be provided with a double linked list of step objects consolidated by a pipeline object. Pipelines provide functions for finding starting points, setting input values etc., while step objects hold dictionaries for both parameters as input values and properties, which can flexibly be used by scheduling algorithms.

Pipelines are provided by dynamically parsing a business process model persisted as XPDL string in the IMDB as described in Chapter 2. The supplied parser has the following restrictions:

- **Disposable**: One parser can be used for one pipeline and becomes depleted,

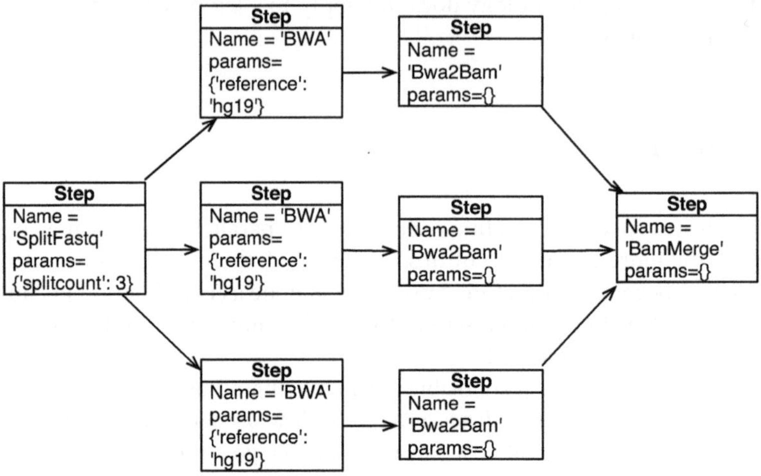

Fig. 3.1: An UML object diagram of a step data structure. Arrows show the order of jobs and the delivery of input and output parameters. Split jobs have multiple subsequent jobs, whereas merge jobs have multiple predecessors. The output of split jobs has to be handed to all jobs of the parallel sub process, whereas merge jobs get all outputs from their multiple predecessors.

- **Database access**: Each parser needs access to the database to fetch the pipeline and the XPDL string, which are the foundation of the parsing process, and
- **Task identifier**: The parser needs the corresponding task identifier of the pipeline to fetch the right data from the database.

In order to simplify tests for different scheduling algorithms, it should be straightforward to swap them. Thus, there should be no need to modify program code.

3.3.2 In-memory Database as Scheduler

An IMDB is used in order to distribute subtasks among workers. All available subtasks are written into one database table, which serves as execution log at the same time. Rows also contain status information and parameters needed for execution, e.g. the path to input files.

An available worker chooses the next subtask from the table. It inserts a new row indicating that this subtask will be executed by it. A unique database index prevents an insertion of an entry from another worker, which chose the same task in the meantime. In this way, only one worker is allowed to process a subtask and consistency is guaranteed by the database.

At the completion of a subtask, a worker adds a row indicating the completion of the task. The same process is applied in case of an error. In this way, all execution information is available for evaluation and analysis within the IMDB.

A basic approach for executing a task is starting subsequent subtasks by the worker, which has finished processing a subtask. This is done by inserting a new row into the subtask database table. Starting a whole new task is as simple as inserting its first subtask into the database table. There is no need for a centralized organization instance except for the database itself.

Nevertheless, there are some disadvantages of this approach. By mixing algorithms to process the data and organizational code, flexibility for reusing jobs for other pipelines is not given: Each job needs to know at least a small part of the pipeline it belongs to. Even the knowledge of merge jobs, which combine parallel execution strands, must be in the worker code itself: a worker should not even take a subtask if it cannot be executed to prevent waiting time. If jobs will be reused in other pipelines, the implementation code of jobs as well as the worker itself has to be adapted. In order to avoid this, centralized scheduling algorithms are discussed in Section 3.3.3.

Thus, in addition to the subtask table, a task table holds their status, a reference to the pipeline they execute, and data affecting the whole pipeline, e.g. parameters.

By exchanging the status of subtasks through the insertion of new rows for every status, the table of subtasks is simultaneously a transaction log for subtask executions. All events like emerging, incurring, finishing, and errors of subtasks persist in the IMDB with all relevant information and form rows in the subtask table. While this approach accumulates more than 150 rows of data per task on average, they can normally be used without much withdraw, e.g. for enabling backup and recoverability in case of a node crash, or as basis for analysis like done by the scheduling algorithm described in Section 3.3.3. This is enabled by the IMDBs powerful storage concepts [78].

Although all features described by them may benefit the convenience of the database in general, some notably facilitate my methodology.

Lightweight compression, as described in Section 1.4.3, reduces memory consumption of redundant information by storing it in a compressed manner in a transparent way. I have experienced a compression rate of 3.5 : 1 for the subtask table. Compressed data is not just smaller, but enables higher density per memory space and therefore alleviates attacking challenges of I/O-bound activities. When reinserting a subtask with just the status field changed, almost no additional memory is used to add this record.

That is, because values of the other columns are already part of the columns' dictionary, mapping from specific, long values to smaller representations. In the data column itself, only the index of the corresponding dictionary entry is saved. Compressed data is not just preferable for better space consumption, but also more suitable for working directly on them. For example, a certain job is done by finding the index of the wanted job and further on this is used for comparisons, joins and filters. Because job names are highly redundant and saved as space consuming VARCHAR column, we can benefit from compression when saving subtasks.

Tables are stored in columnar format, e.g. on disk, records are not saved agglomerated together, but every column forms it own array of values. This aids in two different ways when processing data: First, not whole records have to be read when just few attributes are needed. This exploits obtainable I/O restrictions, because no irrelevant data has to be moved or takes unnecessary place in caches. Secondly, data types with fixed lengths can be saved like an array and be sorted independently, so reading a single or only few values enables directly jumping to the suitable index instead of searching linearly. For example, the work balancer introduced in Section 3.3.3 is only interested in the `subtask` and the `worker_id` columns from the sessions database table and therefore does not need to fetch whole records.

Partitioning is the key for exploiting maximal calculation power without being blocked by locks or the need for interchanging data. When designing proper partitions, only one partition has to be touched by the corresponding queries. Additionally, partitions can be distributed among nodes, so every worker has its own part and can use the database locally without network latency. As a result, response time is improved through IMDB features.

3.3.3 *Real-time Scheduling*

In order to organize execution of diverse GDP pipelines, I propose a single entity to coordinate and schedule execution sequence. I define real-time scheduling as dynamic scheduling, making decisions based on latest relevant data instead of using predefined or pre-calculated values.

Classification of Scheduling Algorithm Requirements

When evolving a scheduling algorithm for GDP pipelines, requirements have to be specified in order to categorize them and find suitable algorithms known from operating systems. Therefore, I first describe some traits of GDP pipeline scheduling and then propose shortest task first as an appropriate solution.

Classification of the environment mainly affects the goals to pursue and therefore determine the set of scheduling algorithms to be chosen from. I argue, the typical requirements of real-time systems do not apply for GDP pipeline execution environments, e.g. the value of processing results does not expire within seconds. However, deciding between goals of batch systems and those of interactive systems is arbitrary. Achieving goals of batch systems benefit the goals of interactive systems and vice versa, e.g. accomplish short response times in average requiring at least some jobs to have a short turnaround time. Still, they compete on a fully stretched system [73, Sect. 9.1]. I concentrate on batch processing goals, because I assume for a fully loaded system high throughput and low turnaround times are of paramount importance in order to overcome peak times quickly.

Furthermore, a decision has to be made whether scheduling is performed based on the subtask or the task level. Task scheduling means deciding first, which task should be executed next and then choose an attendant subtask. Subtask scheduling does not necessarily mean to ignore the corresponding task when coming to a scheduling decision, but choose from the set of all ready subtasks. There are scheduling algorithms fitting for both levels, e.g. FCFS works for both task and subtask scheduling. Still, there are differences in the capabilities: The scheduler described in Section 3.4 does not have the possibility to interrupt workers while they are executing a subtask, so for subtask scheduling just non-preemptive scheduling algorithms will suite. But, when scheduling tasks, every termination of a subtask offers the scheduler the opportunity to plan the subtask from another task first. This means, while traditional preemptive scheduling algorithms do not work without adaption to this constraint for tasks, they still should be considered.

On one hand, scheduling subtasks is promising when focusing on resource utilization, e.g. executing subtask with most subsequent subtasks first in order to enable maximal parallelization. On the other hand, high throughput of subtasks does not necessarily mean producing relevant results for users quickly, e.g. in a bad case solitary subtasks at the end of a pipeline could be postponed repeatedly. Thus, I think optimizing both throughput and turnaround time of tasks are more sensible for users. Therefore, I decide to focus on scheduling tasks. Once again, future work might investigate whether optimizing resource utilization on subtask level would cause unattended good impact on user experience.

Two-stage scheduling, as described in Section 3.2, is outside the scope of this contribution, because workers of my GDP pipeline execution environment, i.e. processors, do not need any more time to swap between subtasks, i.e. threads, from different tasks, i.e. processes, than between subtasks of the same task.

Shortest Task First

In order to achieve maximum throughput, the execution of the shortest available candidate first is the best method, as discussed in Section 3.2. By scheduling on task layer, remaining execution time of all tasks with ready subtasks needs to be estimated. Because the scheduler recalculates estimations whenever a next subtask is requested, risk of falling into problems, caused by unfavorable displaced starts is diminished.

Although scheduling of tasks is neither clearly preemptive nor non-preemptive, by adding up the estimated times of pending subtasks the scheduler can calculate remaining time. Using the count of base pairs in the tasks FASTQ input file does the estimation of process time for subtasks. First, a task with the most similar input size is searched for, then all execution durations of jobs with the same name like the inquiring are aggregated.

With a growing history, estimations will become more and more accurate, because there will be a closer base pair count for each new task then when there are only few tasks with few base pair counts.

Moreover, estimating subtasks that are executed in parallel in the same task becomes more precise during execution of this task: When the first subtask of a parallel part should be executed, there will likely be no close reference values in order to assume a plausible execution time. But, when estimating the second subtask with the same job, execution details of the first will already be in the database. Because they belong to the same task, they have the same base pair count, which leads to proper estimation of the second based on the duration of the first, the estimated time for the third will be the average of both the two forerunners etc.

Traditional large-scale databases are divided in two focuses:

- **Online Transactional Processing (OLTP)** for working with current data at large scale, many short queries, and
- **Online Analytical Processing (OLAP)** for working with pre-aggregated data for fast analysis for few, long running queries [78].

The need of OLAP databases lies in the weak point of fast aggregation in traditional database system, especially when data is continuing to change because of running transactions. When pre-aggregating results, fast analysis can be simulated on costs of real up-to-date information, which is held in the OLTP system.

Combining analysis and transactional operations is not possible in traditional databases in the way IMDBs support this. This is mainly possible because aggregation algorithms are adapted and enhanced to work in parallel on partitioned datasets. When analyzing the execution history of subtasks, it is extremely important to choose suitable partitions of the subtask database table, in order to allow early optimization. I propose to partition using the task id as key in order to distribute among all available nodes. Using multiple processor cores on a single node uses partitioning using the subtasks as the second criteria to exploit parallelization contingency. At the same time, rows with the same subtask are grouped in one partition, which is precondition for early, parallel duration calculation.

Performing calculations on both most recent data and historic data, one rather aggregates all data on the fly then vitiates the results for speed problems. Therefore, neither duration statistics of subtasks nor of tasks are ever saved to the database in order to follow the concept of on-the-fly aggregations to include latest transactional data.

Balancing the Workload

To make use of multiple workers, subtasks have to be spread among computer nodes. A first approach might be to insert all ready subtasks in the database log, so each free worker takes the next one. Indeed, such a greedy way of work balancing prevents the scheduler to decide between all ready subtasks. Once in the database and thereby accessible for all workers to take, subtasks cannot be restrained by the scheduler. Instead, responsibility of choosing the next subtask to execute is handedled by the workers.

Therefore, I propose a work balancer that tracks the nodes' status, e.g. how many workers are running on each of them. This information is persisted in a database table. Knowing all available workers and their status, the work balancer can request an available subtask whenever a worker finishes the execution of his current subtask. This non-greedy approach enables the scheduler to hold back the subtasks as long as possible without detaining the workers too much. Delaying scheduling decisions benefits suitable subtasks, which become ready a short time before the request. When being asked too early, the scheduler has no possibility to decide on them.

If the scheduling decision takes too long, the work balancer might reduce the waiting time for workers at the expense of holding back subtasks: It pre-fetches one available subtask for each busy worker, so the waiting time of a worker for the next task is avoided. This competes with the target to leave waiting subtasks for the scheduler as long as possible.

In order to compound both late scheduling decisions and short waiting time for workers, the work balancer might use time estimation for assigned subtasks, too. Just before estimated completion of a subtask, it requests the next free subtask.

3.4 Application Example

In the following, I describe the scheduling architecture and crucial components.

3.4.1 Architecture

As shown in Figure 3.2, the workers are distributed among a cluster of computer nodes. Multiple worker threads can be executed on the same node. Each node is also part of the landscape of an IMDB. This enables workers to have fast access to the local portion of the database. Every scheduling decision is saved in the database before being executed. This supports the database as a transaction log.

Workers use the User Datagram Protocol (UDP) in order to exchange messages between each other, mostly to inform about certain changes in the database, e.g. the scheduler sends a `wakeup_workers`-signal in order to make the workers look into the database for new subtasks again.

The system running on one node is composed of three parts, as depicted in Figure 3.3: A general part as interface for coordination, a worker part, which is responsible for executing tasks, and a scheduling part, which is unique for the whole landscape and is responsible for handy coordination of the subtasks. The following sections explain them in further detail.

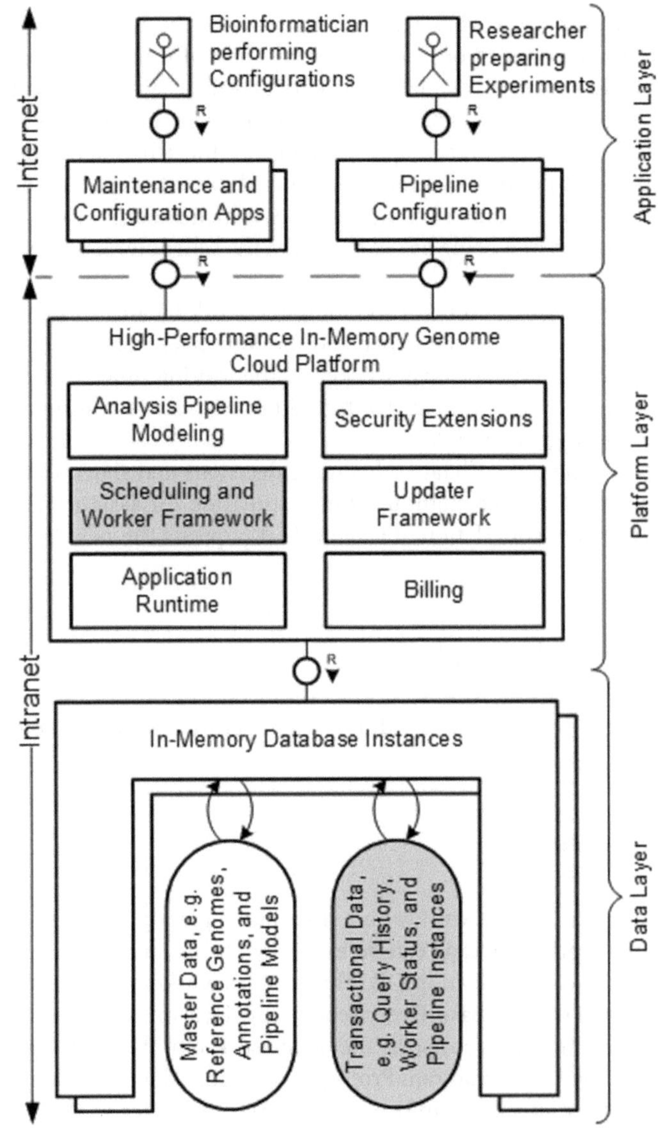

Fig. 3.2: The setup of the distributed worker system modeled as FMC block diagram.

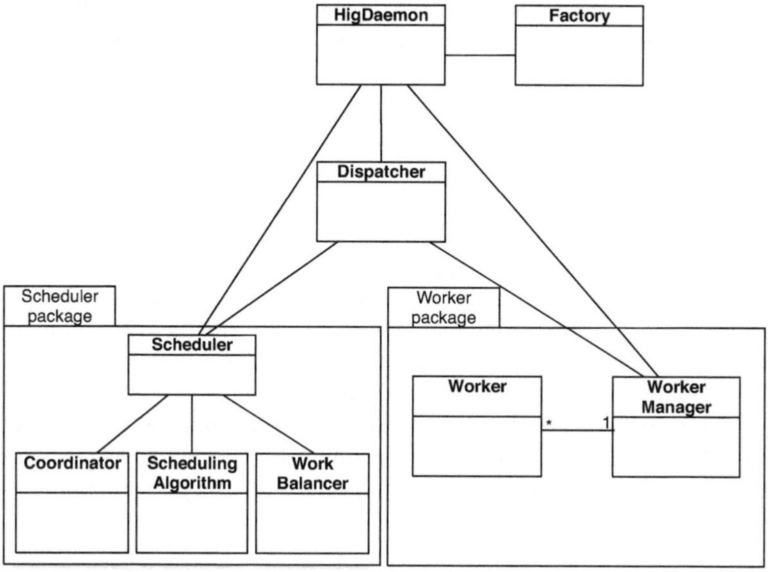

Fig. 3.3: Excerpt of a class diagram created in Unified Modeling Language [68]. The HigDaemon is the main thread, which starts first and coordinates other classes. When acting as worker, a worker manager is used in order to administrate worker processes. The scheduler encapsulates the coordinator, the scheduling algorithm and the work balancer.

HigDaemon, Communication, and Object Creation

On one node, both worker and scheduler are executed as a single process. In order to coordinate them, they are dynamically turned on and off during runtime, which is handled by the `HigDaemon`. It also executes the dispatcher that is responsible for receiving network requests and dispatching them to the respective recipients. Since the dispatcher creates a new thread for every signal and not all classes are thread-safe, compulsory serialization is implemented by the base class of all recipients of the dispatcher.

In order to separate the persistence model from algorithms, there are different repositories managing access to certain parts of the database. While their user classes contain business logic, the repositories hold the Structured Query Language (SQL) code used to communicate with the database. In order to avoid code duplication and to improve code maintainability, all repositories inherit from the same base class, which implements connectivity functionality. This already helped to decouple test code from the database, when writing automated tests for the code.

The factory described in Section 3.4.2 is responsible for creating fully configured objects at runtime. The `HigDaemon` uses it to create nearly all other instances of the system.

Workers for Execution

The worker consists of the actual `Worker` class and a worker manager. The latter is responsible for creating, stopping, and waking up worker processes. Except for wakeup and stop signals from the worker manager, workers are independent. They fetch subtasks from the database in order to execute them by calling the corresponding job module. When there are no more subtasks in the database to fetch, they stop polling and wait for an event of the manager.

Each job is implemented as an own class in its own module. Workers get them by using the job factory introduced in Section 3.4.2 They all have to inherit from the `Job` super class, which provides an interface for the jobs. Selected interface functions are:

- `get_input(key, default = '')`,
- `set_output(key, value)`, and
- `system_command(command, input=None)`.

The first two are needed for exchange of parameters and results of other subtasks. Most jobs just implement a system call with adjusted parameters for the appropriate application that implements their activity.

Coordination, Scheduling, and Work Balancing

The scheduler functions as the mediator of the following classes:

- **Coordinator**, which is responsible for handling step objects and deciding, if a step is ready to execute or not,
- **Scheduling algorithm or policy** that chooses, which of the steps ready to be executed will be the next, and
- **Work balancer**, which dispatches ready subtasks to workers.

All three subclasses hold a reference to the scheduler itself. Although the coordinator will not be exchanged often, both the scheduling algorithm and the work balancer are typical classes to be exchanged in order to meet individual requirements. To decouple their implementations from the scheduler, the scheduler puts the following threefold strategy pattern into action. Each step object is handed from the coordinator to the scheduling algorithm when ready and passed to the work balancer when it should be executed. When workers are available, they notify the work balancer, which pulls a step from the scheduler, comparable to the Kanban productivity system [80]. The scheduling algorithm breaks this pulling chain, because tasks cannot be created just because workers are available. Thus, the coordinator pushes steps as early as they become ready. As a result, the scheduling algorithm has no dependencies and a minimal interface consisting of two methods: `new_step(task_id, step_id, step)` to incorporate a new step and `next_step()` to return the chosen one. The straightforward interface makes developing new scheduling algorithms simple and fast. As described in Section 3.4.2, it also eases interchangeability.

3.4.2 *Application of Design Patterns*

The creation of job instances is not trivial: First, if a requested job is not yet imported, this should be done. Second, some jobs need access to the database, so they need their database connection(s). In order to swap instantiation from the worker class, I used a dedicated abstract job factory, which imports modules of requested jobs from the right package and caches its class object [75]. It also holds parameters required for the job constructor.

The scheduling algorithm determines the next step to be executed. In order to separate the configuration of services from their usage, Martin Fowler proposes two ways, which also achieve simple interchangeability for scheduling algorithms like requested in Section 3.3.1 [74].

Dependency injection, also referred to as inversion of control, proposes to inject the specific scheduling algorithm into the related scheduler. Injection can be implemented using one of the following forms.

- **Constructor Injection**: Dependent classes are injected at construction time. This means they cannot be changed afterwards easily. The constructor will possibly become more complicated, but it will remain the single place to change.
- **Setter Injection**: For every class to be injected, a dedicated setter has to be provided. This means they can be changed afterwards easily, although this is not always desirable.
- **Interface Injection**: For every class to be injected, a dedicated interface has to be implemented by every using class. This is much more invasive than the other approaches.

Fowler also proposes asking a service locator to provide a fully configured instance of the depending class. Using a service locator does not invert control like dependency injection. Each class needs to know which other classes it wants to use, but the service locator handles its configuration. Service locators can specify classes to be initialized during application startup (static service locators) or can make efforts to determine appropriate instances at runtime (dynamic service locators).

I have designed a factory class, which mixes characteristics of both dependency injection managers and service locators. The factory is responsible for creating asked instances with all their dependencies, e.g. at application startup, `factory.get('HigDaemon', node_id = node_id)` is called to create the daemon object. It invokes the factory to import the module from the package where the HigDaemon is implemented. It will then use its configuration to determine the following HigDaemon dependencies.

- **factory**: The HigDaemon will create objects at runtime and therefore depends on the factory, which is passed to it during creation.
- **node_id**: The node identifier is needed for keeping the worker nodes apart from each other. It is provided as a command line argument and just passed through by the factory.
- **session_repository**: In order to log and share its status in the IMDB, the HigDaemon uses a session repository. The factory calls itself recursively to get a valid object for the daemon's constructor.
- **config**: An object representing the configuration of the application. It is created by the factory once and served to any object needing it.

The factory can act as a dependency injection manager, which uses constructor injection, e.g. when creating the session repository with the factory. The HigDaemon uses the factory to create new objects at runtime, e.g. the scheduler or the worker manager. The worker manager also uses the factory to create configured workers on demand. In these cases, the factory is used as a service locator. This enables the coordinator to meet the requirement to use a new parser for every pipeline, as described in Section 3.3.1. Especially maintenance of the system's code base is relieved by use of the factory, because changing dependencies of classes does not involves modifying code from any user. All module imports are made dynamically and centralized, reloading modified configuration and even source code can be done at runtime of the system.

The configuration of the factory is provided as a Python module containing two hash maps: The first one maps class names to their dependencies, the module and the package in which they are located, and if a created object should be cached. It configures the dependency injection part of the factory. The second one configures the service locator part by mapping role names to the classes, which implements them, e.g. *"scheduling_algorithm"* $\Rightarrow$ *"FirstInFirstOut"*. Although these hashes are supplied as Python code, it would be easy to parse them from a JSON string or an XML file like proposed by Martin Fowler [70, 72, 74].

In order to save resources, the factory supports caching on two layers: Class objects are always cached, and instances are just cached when configured as cacheable, e.g. the UdpSender can be cached because it is thread-safe, but there should be multiple workers in the system so their instances should not be reused and therefore not be cached.

3.4.3 Implementations of Scheduling Policies

How the scheduler algorithm decides, which of the steps ready to be executed will be the next, depends on the specific policy implementation. Although I mainly pro-

```
from collections import deque
class FirstComeFirstServed(object):
    def __init__(self):
        super(FirstComeFirstServed, self).__init__()
        self.steps = deque()

    def new_step(self, task_id, step_id, step):
        self.steps.append((task_id, step_id, step))

    def next_step(self):
        return self.steps.popleft()
```

Listing 3.1: First-come first-served on subtask layer. Collections.deque is a faster implementation of a Python queue than a simple list would be [79].

pose shortest task first estimated on the input base pair count, I have implemented several other scheduling algorithms, which may be of certain interest.

The most used one is FCFS on the subtask layer. The complete code is shown in Listing 3.1.

I decided to schedule tasks and not subtasks, as discussed in Section 3.3.3. Therefore, I adapted this principle to tasks by managing a list of task identifiers, which constitute the order of their appearance and a hash for mapping from task identifiers to a list holding the count of ready steps for this task, and a sub-scheduling algorithm, which maintains all steps for this task.

In addition, I implemented another scheduling algorithm categorized as interactive scheduler, as discussed in Section 3.2. It is called lottery scheduling policy, explained by Tanenbaum and introduced by Waldspurger and Weihl [82]. In my implementation, every subtask gets a fix number of tickets. When a subtask finishes, it hands its tickets to its subsequent jobs. Two goals should be achieved with this: Firstly, the more subtasks a task already has processed, the more likely it is for the remaining subtasks to be drawn. This prevents tasks to get stuck immediately before completion. Secondly, parallel jobs are handicapped by splitting receiving tickets with the other jobs, in order to boost non-parallel parts of pipelines until they split again so as to have always enough free subtasks to use all available workers. As drawing lots is not deterministic, however, it is not likely to result in the best decision.

In order to implement the STF policy, as explained in Section 3.3.3, I have used a similar approach to FCFS for tasks: Manage a hash map, which maps task identifiers to sub scheduling algorithms and additional information in order to choose one task from it using additional information. For shortest task first scheduling, this is mainly the estimated remaining time. I have summed up the estimated times of all finished subtasks in order to determine the remaining time for a task.

So as to estimate remaining time of a subtask, I used the IMDB to first find the nearest input size, which belongs to a task containing the same job I am estimating as shown in Listing 3.2.

```
SELECT fastq_readcount, fastq_readcount - :fastq_readcount AS diff
FROM WORKER.TASKS
WHERE fastq_readcount != 0 AND id IN
  (SELECT task FROM WORKER.SUBTASKS WHERE job = :job)
ORDER BY ABS (fastq_readcount - :fastq_readcount) ASC
LIMIT 1
```

Listing 3.2: Identification of tasks with similar input size

```
SELECT AVG(seconds_between(start_log.updated_at, end_log.updated_at))
  AS average_duration
FROM
    (SELECT SUBTASK, UPDATED_AT
        FROM WORKER.SUBTASKS
        WHERE status = 1 and job = :job_name
        AND task IN
            (SELECT id
            FROM WORKER.TASKS
            WHERE fastq_readcount = :fastq_readcount)
        ORDER BY subtask desc) AS start_log
    JOIN
    (SELECT SUBTASK, UPDATED_AT
        FROM WORKER.SUBTASKS
        WHERE status = 2 and job = :job_name
        AND task IN
            (SELECT id
            FROM WORKER.TASKS
            WHERE fastq_readcount = :fastq_readcount)
        ORDER BY subtask desc) AS end_log
     ON start_log.subtask = end_log.subtask
```

Listing 3.3: Calculating the average duration for all same jobs belonging to a task with given input-size. Performing the complete calculation within the database improves response time.

Then, I have taken the average durations of subtasks belonging to similar tasks. Partitioning enables parallel data processing, e.g. to incorporate all available CPU cores. Therefore, I have partitioned both tables using the task identifier in two groups, one for each node of the test system. For the subtask database table, I have used a two-level partitioning, because the subtasks table has around 150 times more rows than the task table and both identifiers are needed in the second query as shown in Listing 3.3. I have chosen the partition count to be the processor count of the test system.

3.5 Evaluation and Discussion

Because the liability for organizing all task executions is pushed toward one node, e.g. the scheduler node, the whole systems becomes vulnerable to faults caused by this node. In order to minimize potential outages, high-availability features should be applied. This includes dynamic takeover of the scheduler role by another node in case of the scheduler node's crash, inaccessibility, or overloading.

I propose to implement this by determining two or more slave schedulers, which a) check response time of scheduler periodically, and b) receive and process the same events sent to master scheduler just without releasing steps [76]. Slave schedulers follow a exchangeable fail-over policy in order to decide, when the master is considered to be unreliable. If necessary, they will take over the master role and use a database constraint to prevent creating multiple masters. Due to redundant event handling, the new master scheduler can take over without any delay.

3.6 Conclusion and Outlook

Results of GDP pipelines are the basis for much information physicians or researchers need. Therefore, enhancing them to be faster and more flexible is the foundation for broadly using genomic data for research and medical treatments.

I proposed a cluster of worker nodes using an in-memory database, coordinated by a single scheduler to achieve these goals. I especially have described, how dynamic execution of pipelines can be implemented. Furthermore I have explained basics of scheduling, how this can be adapted to GDP pipeline scheduling, and how I implemented some simple interchangeable scheduling algorithms. I illustrated how shortest task first scheduling maximizes throughput, how it is dependent from the IMDB to be fast in analyzing execution logs, and how it behaves when used on huge amounts of underlying data. Although it is not the fastest on relatively small history, is seems to grow just with logarithmic complexity towards the size of the history table.

I also have discussed how the system can stay fault-tolerant although it is highly dependent from one instance to take the scheduler role.

Future work could prefer goals of interactive systems in order to improve user experience during times of low workload, or might investigate if optimizing resource utilization by scheduling on subtask layer causes unattended good impact on user experience or task throughput.

Also, an implementation of the aforementioned two-stage scheduling could benefit from locally saved intermediary results – a feature that is still missing in the research prototype.

Lastly, future work could exploit how the scheduler could handle a cluster of heterogeneous worker machines, e.g. when some jobs can just be executed on a subset of the worker machines.

3.7 References

[68] Ambler SW (2005) The Elements of UML 2.0 Style. Cambridge University Press

[69] Ansorge WJ (2009) Next-generation DNA Sequencing Techniques. New Biotechnology 25(4):195–203

[70] Bray T et al. (1997) Extensible Markup Language (XML). World Wide Web Journal 2(4):27–66

[71] Cock PJ et al. (2010) The Sanger FASTQ File Format for Sequences with Quality Scores, and the Solexa/Illumina FASTQ Variants. Nucleic Acids Research 38(6):1767–1771

[72] Crockford D (2006) RFC4627: The application/json Media Type for JavaScript Object Notation (JSON). http://www.ietf.org/rfc/rfc4627.txt. Accessed Sep 23, 2013

[73] Ehses E et al. (2005) Betriebssysteme - Ein Lehrbuch mit Übungen zur Systemprogrammierung in UNIX/Linux. Pearson Studium

[74] Fowler M (2004) Inversion of Control Containers and the Dependency Injection Pattern. http://www.martinfowler.com/articles/injection.html. Accessed Sep 23, 2013

[75] Gamma E et al. (2001) Design Patterns: Abstraction and Reuse of Object-oriented Design. Springer

[76] Li Ff, Yu Xz, Wu G (2009) Design and Implementation of High Availability Distributed System based on Multi-level Heartbeat Protocol. In: Proceedings of the International Conference on Control, Automation and Systems Engineering, IEEE, pp 83–87

[77] Li H, Ruan J, Durbin R (2008) Mapping Short DNA Sequencing Reads and Calling Variants using Mapping Quality Scores. Genome Research 18(11):1851–1858

[78] Plattner H (2013) A Course in In-Memory Data Management: The Inner Mechanics of In-Memory Databases. Springer

[79] Python Software Foundation (2013) Using Lists as Queues. http://docs.python.org/2/tutorial/datastructures.html. Accessed Sep 23, 2013

[80] Sugimori Y et al. (1977) Toyota Production System and Kanban System Materialization of Just-in-time and Respect-for-human System. The International Journal of Production Research 15(6):553–564

[81] Tanenbaum AS (2009) Modern Operating Systems, 3rd edn. Pearson Prentice Hall

[82] Waldspurger CA, Weihl WE (1994) Lottery scheduling: Flexible proportional-share resource management. In: Proceedings of the Operating Systems Design and Implementation, USENIX Association

Chapter 4
Exchanging Medical Knowledge

Franz Liedke

Abstract Genome data exists in many different formats in distributed data sources. Due to the many possible uses, various tools have been created for analysis and visualization. The huge size of the data sets combined with limited network capacity makes it very time-consuming to use cloud applications. In this contribution, I propose to transfer application logic to the data instead. This can be achieved by making the apps work independently from how and where the data is stored. In-memory databases enable this through on-the-fly data transformations and specialized query execution plans for distributed data.

4.1 Introduction

Innovative medical knowledge, both data and tools, should not be withheld, but shared with researchers, institutes, and physicians to improve medical research. The idea of exchanging data – made popular by Web 2.0 applications – has arrived in medicine and many tools that encourage sharing of knowledge have been built [90, 92]. This kind of sharing is especially important in the context of human genome research [100].

Mutations in the genome are often a root cause for diseases [108]. However, there is an abundance of different mutations that can indicate the same superficial result, e.g. the same disease [94]. Gaining an understanding of these processes is only possible with large amounts of data – combining the genomes of large numbers of human beings. Thus, research institutions in the genome field depend on having access to large data sets. Withholding knowledge in this context would decrease the efficiency of conducted research [83].

Sequenced genomes are usually stored in the FASTQ file format [89]. The sequencing results are commonly read multiple times to increase quality. Uncompressed and with thirty-fold coverage, the approx. 3.2 billion base pairs of the human DNA, along with a confidence value for each base, can take up between 100 and 200 GB of storage. When dealing with data of this size, data handling becomes a complex task [107]. For example, uploading these files to websites consumes minutes to hours.

Another problem with sharing medical data is the heterogeneity of the existing storage formats. Various file formats exist and there is only a limited standardization. To enable applications to work together, a common suggestion is standardization. As multiple apps use similar data, a common standard data format is required. However, standards need to be well defined and suitable for multiple use-cases. Also, existing data has to be converted, which means that converters have to be implemented, tested and finally executed on the huge amounts of existing genome data. Nonetheless, attempts are being made to create standard definitions for the different kinds of data in medical care and research. These efforts are necessary, but it will take some time until these standards are widely known and used.

In the meantime, analysis tools should still be able to work with as many different storage formats as possible. What if we could execute necessary tools directly on the required data without transferring them? Instead of fitting all existing data to a new standard, we can try to make the solutions work despite the heterogeneity.

In this work, I demonstrate how IMDBs can be used to share both medical data and tools operating on them. Firstly, existing data can be mapped on the fly to the format expected by applications to avoid the standardization problem. Additionally, IMDBs are able to handle enormous amounts of data. Because of their capability to distribute work across nodes, they can also be used to efficiently run calculations on heterogeneous data stored in distributed locations.

The remainder of this work is structured as follows. Section 4.2 deals with previous work in the fields of sharing medical knowledge and the technical backgrounds of useful IMDB features. In Section 4.3, I will describe the structure of my research prototype. The setup and results of the benchmarks testing the performance of IMDBs running queries on distributed data will be explained in Section 4.4. These results will be discussed in Section 4.5, followed by a conclusion and a look at possible future work in Section 4.6.

4.2 Related Work

In this section, I will explain selected key benefits of IMDBs as well as reference a number of signification approaches to enable this kind of sharing.

4.2.1 In-memory Databases

By operating entirely on main memory, IMDBs are extremely fast compared to traditional database management systems [104]. In this section, I will describe selected features, which my work benefits from.

Non-materialized Database Views

Materializing database views is a common optimization technique in disk-based databases. As explained by Gupta and Mumick, this is mainly done to gain performance [95]. The materialization of views increases performance as it avoids recalculation on every access. The speed of access can be further increased through the use of indices, which can now be built on top of the database views.

However, according to Gupta and Mumick it comes with the disadvantage that the data is stored redundantly – in its original form and after transformation –, wasting space on the hard drive. This leads directly to the second problem, their maintenance: if the underlying data changes, the view must be updated accordingly. In some cases, this means the entire view needs to be regenerated, which can be quite expensive. In other cases, only parts of the views are updated. Detecting the underlying changes and reacting to them is called view maintenance. The usefulness of various view maintenance algorithms are discussed in detail by Gupta and Mumick.

IMDBs are fast enough to recompute views on the fly [106]. As the views are generated anew in-memory every time they are needed, the aforementioned disadvantages – redundancy and maintenance – can be avoided [103].

Parallelization

IMDBs take advantage of parallelization by distributing the workload across different CPU cores of a server and across multiple compute nodes. Parallelization can be applied at different levels inside a database, e.g. inter- and intra-transactional and on query and operator level [103].

The strategies used for parallelization depend on different factors such as the architecture of the system. In a shared-memory architecture, all the data is still available for each processor. In a shared-nothing system, the data needs to be moved to executing processors. Thus, it is advisable to send only queries and their results over the network instead of transferring intermediate results or all of the data in order to run calculations on one processor accessing all the data. Specialized query plans exist to optimize scan, aggregation, and join operations for such a distributed architecture. As a result, the benefits of inter-operation parallelization are available when distributing data of the same type across many nodes, i.e. horizontal partitioning as described in Section 1.4.4 [85].

4.2.2 Sharing Medical Knowledge

Research on how to share medical knowledge is constantly in research focus. In 1988, Clayton et al. stressed the importance of enabling institutions to share the contents of their medical knowledge bases. They observed that "[...] a major problem is the fragmentary and incomplete nature of this knowledge base [87]." Furthermore, they discussed difficulties and obstacles for data sharing such as liability issues and the actual willingness to share. The suggested approach to sharing involved a common representation of knowledge base contents. The authors concluded that "[...] knowledge would be distributed in modular [output] suitable both for direct scrutiny by medical personnel [...] and for translation to one of the supported systems." They explained that such a format would need to be understood by human beings and parsed by machines, while being flexible enough to be translated to any of the heterogeneous knowledge base systems.

One year later, the same authors explained in more detail what their solution could look like [88]. They also described how logic stored in knowledge bases could be shared and prepared for mixed interaction, as opposed to sharing only "passive modular knowledge base contents." They examined a SQL-like syntax as example. Finally, Clayton et al. made it clear that encouraging the concerned parties to share knowledge could also mean enabling them to make money in the process: "Sharing does not necessarily mean that the knowledge which an institution or group has laboriously translated to a computer representation should be given away gratis, but does mean that the work of one group can be transferred to another environment [88]."

The general method of Clayton et al. – finding a common representation of knowledge and using that to distribute data between the different systems – was also used in several other publications. A selection of them will be evaluated in the following.

The Secondary Use of EHR Data Project

The Secondary Use of EHR Data (SHARPn) Project of the Strategic Health IT Advanced Research Projects (SHARP) Program is investigating how the Electronic Health Record (EHR) can be used in secondary applications such as research [86]. Rea et al. declare standardization as the main problem of data exchange: "One of the primary informatics problem areas in this endeavor is the standardization of disparate health data from the [...] many health care organizations and providers [105]."

The SHARPn Project is split into six separate but intertwined sub-projects, e.g. one is working on "Clinical Data Normalization Services and Pipelines". Their normalization solution consists of two tenants. First, they rely on a "common information model" for declaring standard formats for medical entities. They use the Clinical Element Model (CEM), which describes the formats and can be compiled into various other representations, e.g. programming language classes or XML schemes [102]. "Terminology mapping services" are then responsible for mapping the concrete local information to the CEMs, using standard terminology.

According to Rea et al., the SHARPn Project's advantage over traditional applications is that tools using the platform are not "[...] limited to only the specific use cases for which they were developed [105]." At the same time, they acknowledge the "[...] challenge to create canonical models of EHR data for secondary use, as there are a variety of purposes to be served." In the light of this challenge, they discover the need to handle multiple terminology versions and their updates.

BaseSpace Cloud Computing

A solution more aimed towards genome sequencing data is the BaseSpace cloud computing platform [97]. Developed by Illumina, Inc., sequenced data from their own sequencing machines can be streamed directly to the platform via the Internet. The standardized data in this large accumulated data store can then be accessed by custom apps for analyzing the data. The service is backed by Amazon Elastic Compute Cloud (EC2) and thus provided with the power to scale and handle the huge amounts of information. Apps can access the data from the platform through a web API and can be distributed across the platform.

Further Approaches

There are other approaches to enable the sharing of knowledge across medical systems. For example, Giuse et al. state that "[...] there is relatively little sharing and reuse of individual components" of health information systems [91].

The need for integration of data from multiple sources has also been identified as one of the main challenges in today's bioinformatics by Molidor et al. [101]. It resulted in a call to "[...] introduce standardization of terms and data formats." The authors also emphasized the need for high-performance technologies that can speed up analysis of genome data, comparing the growth of entries in the GenBank database to the ever-increasing packing density of transistors on computer chips.

All of these methods have the request for standards for representation of medical data, along with the ability to convert differently formatted data into these standard formats, in common. XML is often used to describe the standards [84, 99].

4.2.3 Requirements

Medical knowledge bases grow in size over time. As knowledge increases slowly, the amount of new data that must be dealt with is limited. Transferring that textual data between the different knowledge bases might take a while, but is certainly reasonable. The size of genomes, however, is even bigger. There is a never-decreasing supply of new patients. Thus, there would be a constant stream of new diagnostic data, which needs to be transferred to centralized services.

Application developers need stable interfaces to access the data. Standards are a solution for this problem. The BaseSpace platform tries to enforce such generic formats. However, industry-wide standards are hard to establish and take a long time until properly propagated. Since applications know what kind of information they access, we can let them specify the standards the input has to adhere to. This works very well with IMDBs, as they are able to map large amounts of information on the fly.

Transferring algorithms instead of data appears to be a more appropriate strategy in the context of big genome data. Since IMDBs are also capable to work efficiently with distributed data sets, they are the technical basis for a working implementation. This makes it possible for institutions to open up their data to outside usage, while retaining ownership of their data.

4.3 Application Example

As a research prototype, I implemented a plug-in architecture on top of the High-Performance In-Memory Genome (HIG) project as introduced in Section 1.5. The prototype will be described in this chapter.

4.3.1 Use Case

I defined a software lifecycle process for institutions and individuals to share both their medical data and the applications used to e.g. analyze that data. A concrete process instance as shown in Figure 4.1 consists of the following steps.

1. An application developer **creates** a tool that allows users to analyze the mutations of various patients' genomes.
2. The developer packages the tool and **uploads** it to a centralized repository where users interested in such a program can find it.
3. A research institution has access to anonymized patient genomes from a number of partner organizations, such as health care providers. They **share** their data by creating a landscape of connected IMDBs.
4. The researcher **downloads** the tool as compressed archive file.
5. The tool is **installed** on the local system in the researcher's institute.
6. A bioinformatician working at the institution **configures** the application by mapping the different formats of patient data across the connected systems to a single format as expected by the app.
7. The app is then ready to **run** on the system, using the data from all of the health care providers as input.

4.3.2 System Requirements

The following requirements build the basis for my prototype.

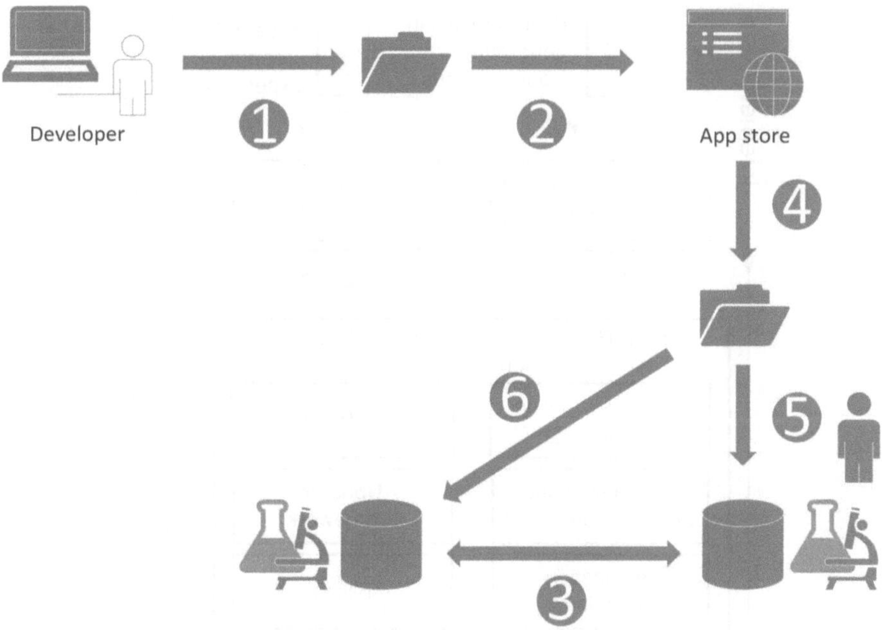

Fig. 4.1: Application lifecycle: From application developer to end user.

- **Extensibility**: Applications can extend central parts of the system.
- **Isolation**: Applications need to be isolated from each other. This is important on the database layer, where apps can maintain individual data structures. They should also be able to access system data they need to operate properly.
- **Pluggability**: Applications can be added and removed without affecting the state of the remaining system.

Extensibility

As shown in Figure 4.2, apps extend the HIG platform at three places: the data layer, the web service and the frontend.

With IMDBs, it makes sense to push down application logic into the database, since calculations, algorithms and business operations can be executed directly on the data, e.g. with stored procedures [103]. The webservice calls these procedures and prepares their outputs for presentation by the actual frontend.

Operating directly on the data has the advantage that the overhead of transporting rows of data from and to the place where calculations are executed is removed. Also, when implementing operations on top of the data in its atomic form in the database, these operations can gain from the speed of data access in IMDBs.

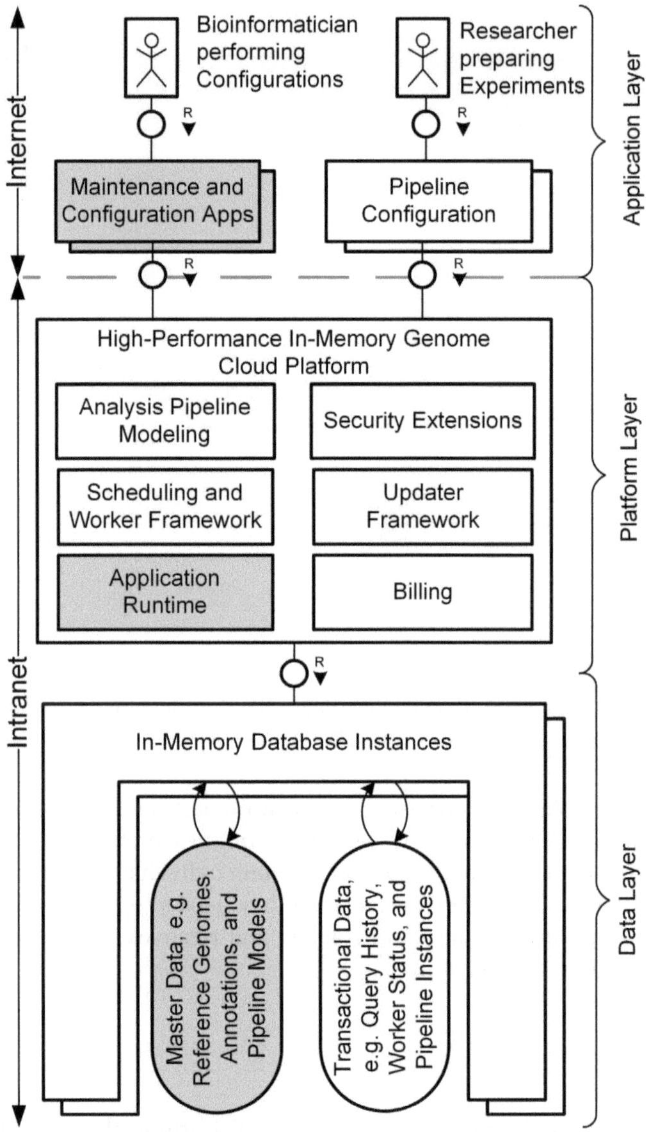

Fig. 4.2: The architecture of the HIG platform modeled with FMC [98].

For our apps this means that both the frontend and the web service should be as lightweight as possible when it comes to logic, focusing only on the presentation of the data, which is retrieved either directly from tables or views, or the results obtained from the execution of stored procedures.

Additionally, individual frontends can be exchanged as described in Figure 4.3 due to the web service, which provides access to the data. To prove this, I built a tablet application as an alternative UI for the cohort analysis app that is described in Section 4.3.5.

Isolation

During installation of applications, a separate database schema and user are created. This user has full permission in that schema, but no permissions outside of it. That way, their view of the database is completely isolated from other apps, which is ensured by the database access control mechanism.

Access to data that serves as input for the app is provided via the views that are generated during app initialization. Since they are created in the database schema of the app itself, read-only access is allowed.

Database access in the web service is provided through a separate connection for each app, using the credentials of their own user.

Pluggability

Apps are distributed as compressed archives and unpacked during installation. Deleting the extracted folder as well as the database schema and user that were created for them uninstalls them. The corresponding row is removed from the database table that holds information about the installed apps, along with their meta information, e.g. database schema and user.

4.3.3 Installing Applications

During installation, a number of SQL scripts can be executed to create the required structure of database tables, views and procedures. Additionally, applications can provide HTML and JavaScript files that can be added as tabs to the web interface of the HIG platform. In order to make data from the database available in the frontend, controllers can also be added to the web service running Ruby on Rails. All of this is as simple as storing a file in a specific location of the app's archive. SQL scripts in the `installation` directory are handled during installation, HTML and JavaScript are stored in the `frontend` folder, and Ruby on Rails controllers are routed directly to the `controllers` directory.

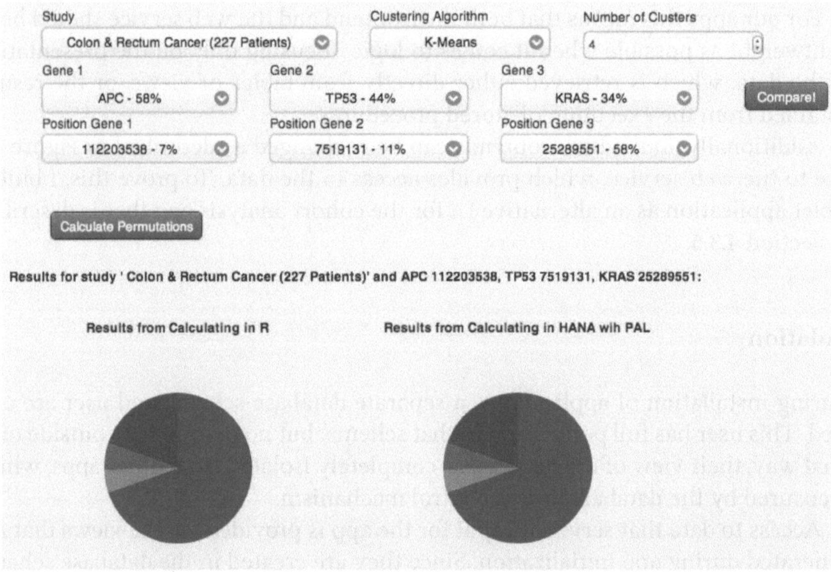

(a) Cohort analysis website interface on a desktop device.

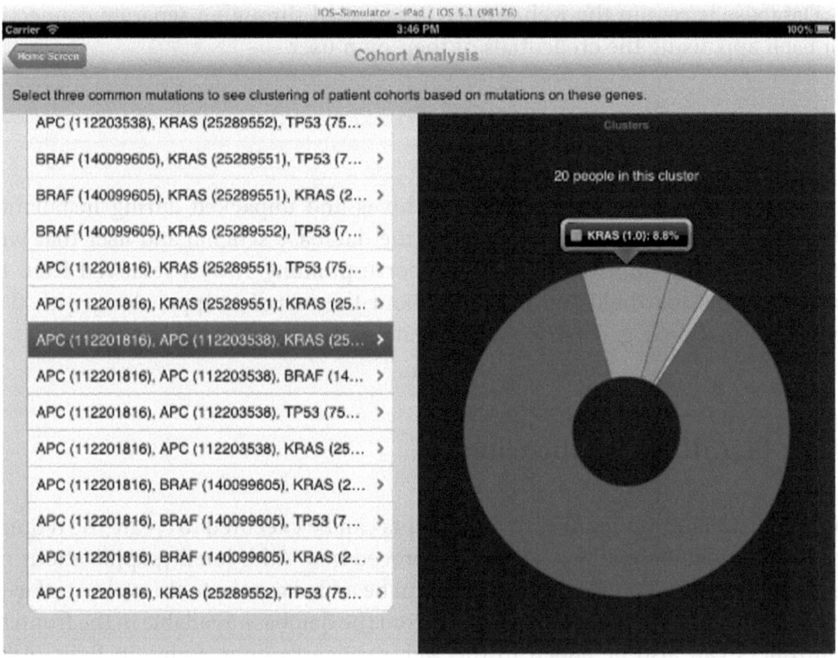

(b) Cohort analysis user interface on a mobile device.

Fig. 4.3: Alternative frontends for the same app.

```
{
  "id": "cohorts",
  "title": "Cohort Analysis",
  "version": "0.1.0",
  "author": {
    "name": "Franz Liedke",
    "email": "franz.liedke@student.hpi.uni-potsdam.de"
  },
  "tabs": [
    {
      "view": "cohorts",
      "name": "Cohort Analysis"
    }
  ],
  "input": {
    "mutations": {
      "patient_id": "string",
      "gene": "string",
      "start_position": "integer",
      "end_position": "integer"
    }
  }
}
```

Listing 4.1: The manifest file defines metadata like author and version for an application, along with tabs to add to the UI and the required format of the input data.

4.3.4 Configuring Applications

Generally speaking, applications operate on entities, which are in turn made up of certain well-defined attributes. In the genome data context, entities are, for example, genes, patients, and mutations. I created a standardized format for manifest files where application developers can specify metadata about their tools, and define what format (name and type) input data has to have, as shown in Listing 4.1.

For apps to work, they must be able to access existing data. Database views can serve this purpose. I developed a tool that can be used to interactively create views that map the input to the desired format without the need for specific SQL code.

For every field of every table specified in the `input` section of the manifest file, the user has to select a matching database field. Automatic completion for field names assists in that process as depicted in Figure 4.4. One-to-one mappings are simple to create; for more complex mappings, SQL functions, such as `CONCAT()` or `SUBSTR()`, can be used.

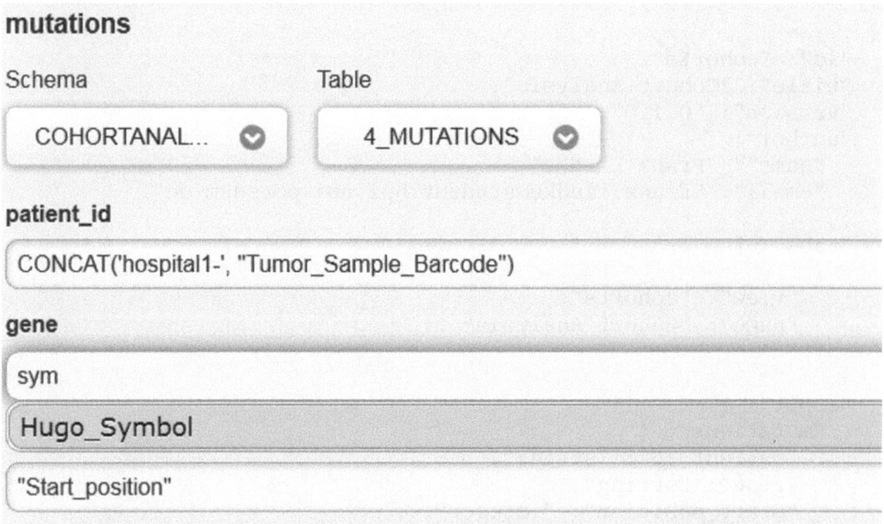

Fig. 4.4: Aided user controls: Fields from the table that match the input are automatically displayed to support the user in selecting the correct ones.

4.3.5 Cohort Analysis

As a proof of concept, I bundled the cohort analysis application out of the cohort analysis application as described in Chapter 6, which allows for the analysis of patient cohorts by mutations on their genes. The user can select multiple genes. Using various clustering algorithms, patients are then divided into several groups, based on which of the selected genes exhibit mutations in their genotype. The tool is split up into three components:

- A database backend, which takes care of the clustering algorithms,
- A controller for the web service that calls stored procedures in the database and transforms it to JSON output, and
- One of two alternative user interfaces as shown in Figure 4.3.

The only input data the application needs is information about the patients' mutations: For each patient, it wants to know an identifier, the mutated gene and the exact position of the mutation. With my configuration tool, it is then possible to use this application either with annotated data from a cancer study, or – through more advanced transformations – with the analyzed genomes of existing patients.

4.4 Benchmarks

The second part of my work is the exchange of large medical data between distributed systems.

In this chapter, I will describe the tests and benchmarks I ran to demonstrate that IMDBs can indeed be used for processing distributed heterogeneous data. I wanted to show that the amount of data sent over the network is indeed limited due to calculations being done where the data is stored. Thus, I tested how certain queries were affected by controlling the network latency.

4.4.1 Method

This section will describe how the benchmarks were prepared, explaining their setup in terms of hardware and data layout, as well as which queries were run.

For benchmarking, I installed an in-memory database instance on two server nodes configured to run in landscape mode as summarized in Table 4.1. I used co-

Processor	8 x Intel® Xeon® E7-8870 @ 2.40 GHz
Main Memory	64 x 32 GB DDR3 1066 MHz + 64 x 16 GB DDR3 1066 MHz
Operating System	SUSE Linux Enterprise Server 11.2 (Kernel 3.0.13-0.27-default)

Table 4.1: Configuration of benchmark system.

hort data as described in Chapter 6 to create two heterogeneous sets of patient data for the purpose of the benchmark: one data set with patient data stored American and the other one with patient data stored in Europe. Both of them contained the same information about patients – an identifier, their year of birth and the name of the gene with a mutation –, albeit formatted differently. The tables can be seen in Figure 4.5.

In order to demonstrate that distributed query plans work even with data that is transformed using views, I created such views to map the heterogeneous data to a common format. For the European dataset, only the year of birth had to be extracted from the birthday using a string operation, while the American patients' data had to be joined with a mapping table to get the name of the mutated gene based on the position of each mutation. A third view combines the two views so that all patients' data can be used together transparently from the outside as shown in Figure 4.6.

To enable realistic sample queries, I also used clinical trial data as described in Chapter 8. I copied a part of this data to a separate schema, which is depicted in Figure 4.7.

The following queries were run during the benchmarks, based on these three types of parallelized algorithms:

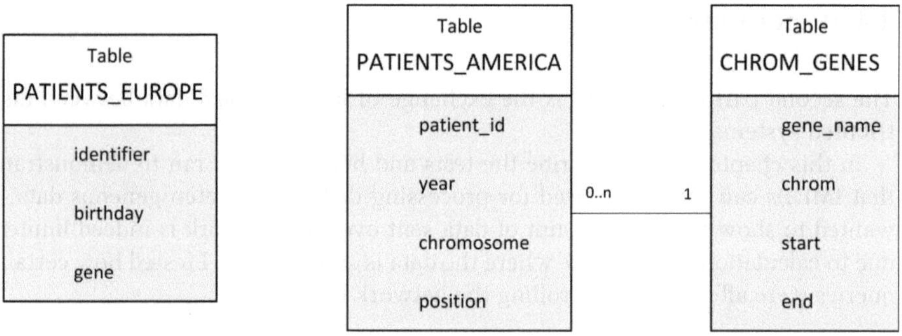

Fig. 4.5: The heterogeneous structure of patient data in the two systems, modeled as UML database diagram [93]. Both database tables with patient data are mapped to a common data format using database views as described in Figure 4.6.

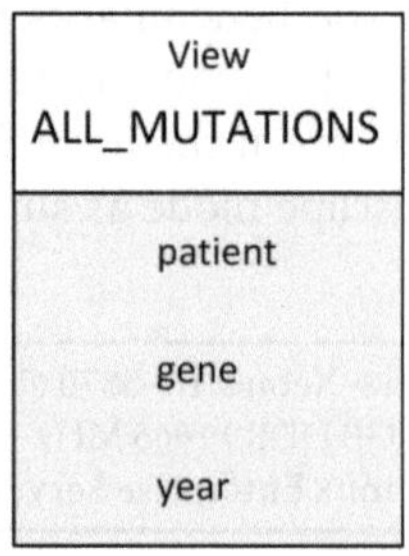

Fig. 4.6: The common format for patient data after transforming with views, modeled as UML database diagram [93]. The two unified datasets are combined using the `UNION ALL` SQL statement in another view. The patient identifiers from both sources are known to be disjoint.

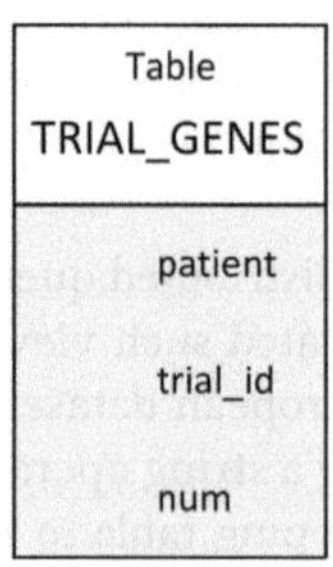

Fig. 4.7: A table connecting clinical trials and gene specified therein, modeled as UML database diagram [93]. The `num` column describes how often the gene name is mentioned in the trial's description text.

```
SELECT COUNT(*) FROM all_mutations WHERE gene = 'KRAS'
```

Listing 4.2: Scan operation: Count the number of mutations on the KRAS gene.

```
SELECT gene, COUNT(*) FROM all_mutations GROUP BY gene
```

Listing 4.3: Aggregation operation: Determine how often genes were affected by mutations.

```
SELECT DISTINCT patient, trial_id, num
FROM all_mutations
INNER JOIN trial_genes
    ON all_mutations.gene = trial_genes.gene
WHERE year = 1973
ORDER BY num DESC
```

Listing 4.4: Join operation: Find all clinical trials that might be relevant for patients that were born in 1973, based on whether their mutated genes are mentioned in the trial description.

```
tc qdisc add dev eth0 root netem delay 150ms
```

Listing 4.5: Defining a minimum network latency of 150 ms

1. **Predicate scan**: A condition has to be evaluated for one column of one table,
2. **Aggregation**: Groups have to be built and aggregations run for each group, and
3. **Join**: Tables are joined across nodes.

Listing 4.2 shows a scan operation to get the number of patients' mutations on the KRAS gene. This query or variants thereof could be run to check whether further investigations regarding this gene could be viable for a given group of patients.

The second query was an aggregation counting how often certain genes were affected by mutations across all patients ad shown in Listing 4.3. This can be used to identify patients with common mutations from a cohort.

The third query combines data from different sources to gain new insights. By joining the list of mutated genes with trial information, the set of patients that certain clinical trials will work for can be identified. The query can be seen in Listing 4.4.

To simulate long-distance connections between both database instances as defined in Figure 4.8, I used the UNIX command line tool `tc` [96]. When running on a shell, it controls the network latency, e.g. to delay it by a well-defined time like 150 ms, as shown in Listing 4.5.

I ran each query multiple times with varying network latency. To get statistically relevant results, for each delay I executed the queries until the Standard

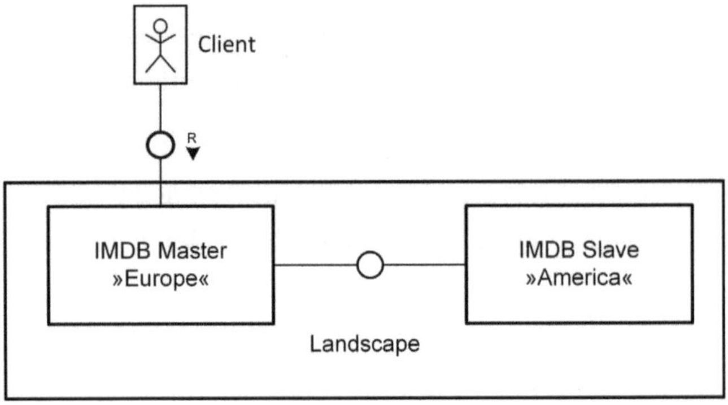

Fig. 4.8: The benchmark setup modeled as FMC block diagram [98]. The network latency between the two databases is controlled, e.g. to simulate a connection between Europe and North America.

Deviation (SD) of the measured times dropped below five percent after applying a confidence interval of 96 percent to catch outliers.

4.4.2 Results

While the relative SDs in the second and third benchmark, as depicted in Table 4.3 and Table 4.4 respectively, are significantly higher than those in the first one, which can be seen in Table 4.2, they are still mainly smaller than 10 % and always below 15 %. This is an acceptable range, as it does not suggest that the test environment was unreliable.

In addition to the mean roundtrip time, these tables also list a normalized roundtrip time. The normalization is obtained by subtracting the latency once for each package that is sent. The number of packages can be derived from the slope of the mean values. These were 1, 11, and 5 for the scan, aggregation, and join operation respectively.

4.5 Evaluation and Discussion

In the following, I evaluate and discuss the obtained benchmark results.

Latency [ms]	MRT [ms]	NRT [ms]	SD [%]
0	15	15	5.8
50	66	16	1.6
100	116	16	1.0
150	166	16	0.9
200	216	16	0.5
250	266	16	0.5
300	316	16	0.4
350	366	16	0.3
400	416	16	0.2

Table 4.2: Execution statistics of the predicate scan operation (MRT = Mean roundtrip time, NRT = Normalized roundtrip time, SD = Standard deviation).

Latency [ms]	MRT [ms]	NRT [ms]	SD [%]
0	76	76	6.8
50	705	155	11.8
100	1,266	166	9.6
150	1,900	250	11.0
200	2,443	243	9.1
250	3,010	260	8.7
300	3,561	261	7.4
350	4,070	220	5.2
400	4,648	248	4.9

Table 4.3: Execution statistics of the aggregation operation (MRT = Mean roundtrip time, NRT = Normalized roundtrip time, SD = Standard deviation).

Latency [ms]	MRT [ms]	NRT [ms]	SD [%]
0	80	80	3.1
50	326	76	13.7
100	606	106	14.1
150	820	70	5.0
200	1,130	130	11.4
250	1,377	127	9.9
300	1,613	113	7.6
350	1,867	117	5.7
400	2,120	120	4.7

Table 4.4: Execution statistics of the join operation (MRT = Mean roundtrip time, NRT = Normalized roundtrip time, SD = Standard deviation)

4.5.1 Predicate Scan

The results of the predicate scan show that the normalized response time remains constant while the network latency keeps growing as depicted in Figure 4.9.

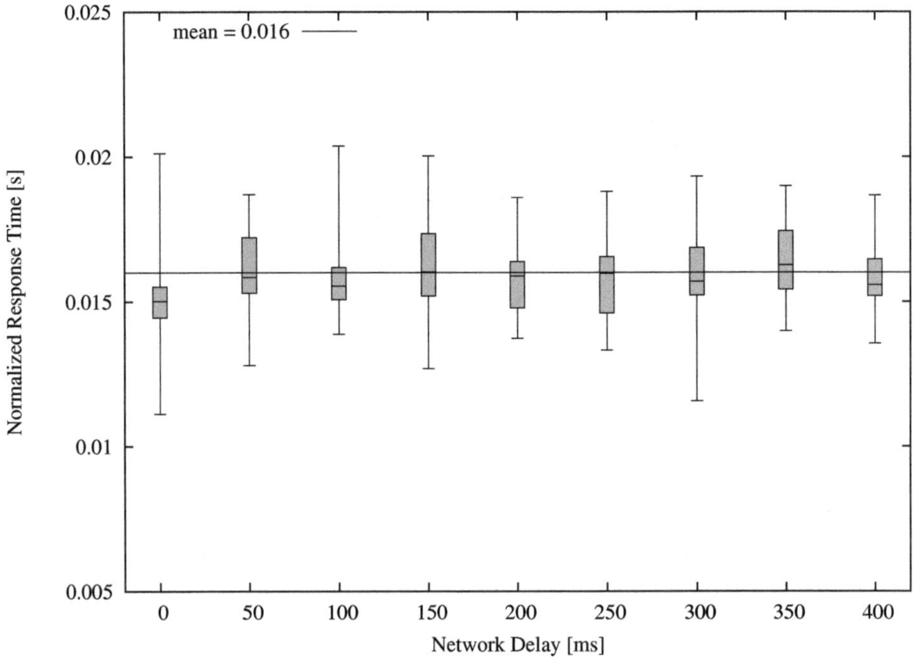

Fig. 4.9: Development of query times of the predicate scan.

This indicates that the query itself is executed in 16 ms while the remaining execution time is spent in a single request response cycle for sending the query and receiving the results via the network.

To evaluate why the overall query processing time is determined by the network delay, we can take a look at the query plan that was created by the IMDB as depicted in Figure 4.10. It shows that the database performs all operations as close as possible to the location of the data. The selections, i.e. filtering by gene name, are executed directly on the patient data. For the `European` dataset this is a simple column lookup, for the `American` counterpart this means that a join with the chromosome-to-gene mapping table has to be carried out.

Exactly one row is sent across the network. It contains the count of mutations on the KRAS gene in America. In fact, the IMDB even parallelizes the counting of the rows by counting the rows in each partition and then adding up the result.

I conclude that the IMDB is able to treat heterogeneous data as if partitioned with the built-in database feature. Even though there was no explicit partitioning,

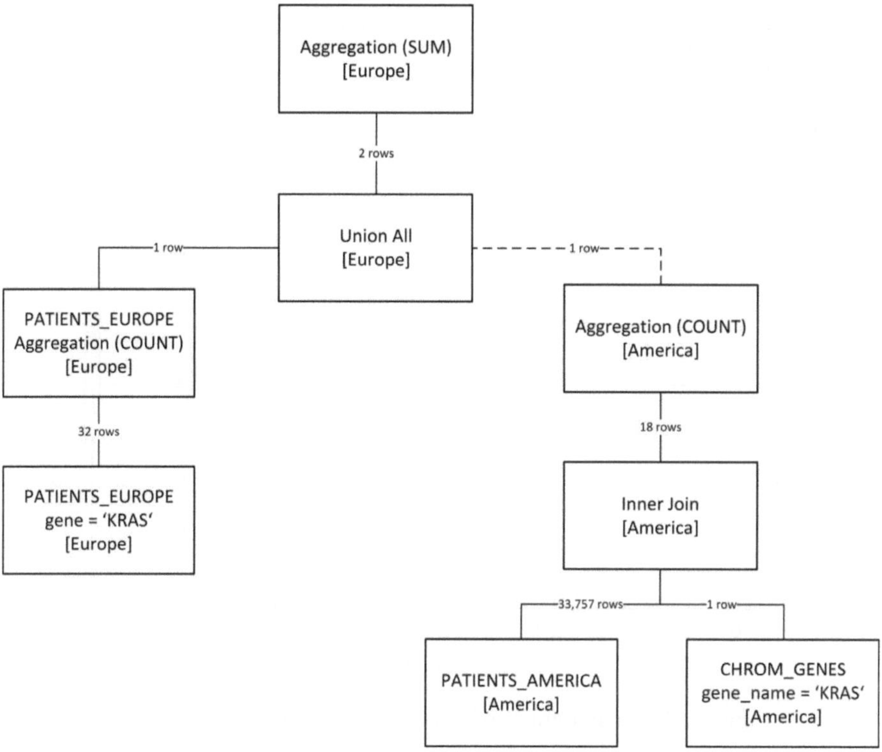

Fig. 4.10: The query execution plan for the scan operation. This query returns one row. Selections are executed in a distributed manner, limiting the number of rows sent across the wire to one.

the database derived a suitable query execution plan from the definition of the view that combined the two datasets using the `UNION ALL` statement.

4.5.2 Aggregation

After normalizing the response time, the aggregation operation shows a similar behaviour as the predicate scan. However, it takes considerably longer to execute in total and the results are more spread out, although the bulk of the measured values is located around the center, which is depicted in Figure 4.11.

The relatively large distribution of the values reflects a varying number of network packets sent during individual benchmarks for this query. I suspect that these variations are caused by fluctuating execution times of internally parallelized pipeline operations, leading to a varying number of intermediate result packets to be exchanged.

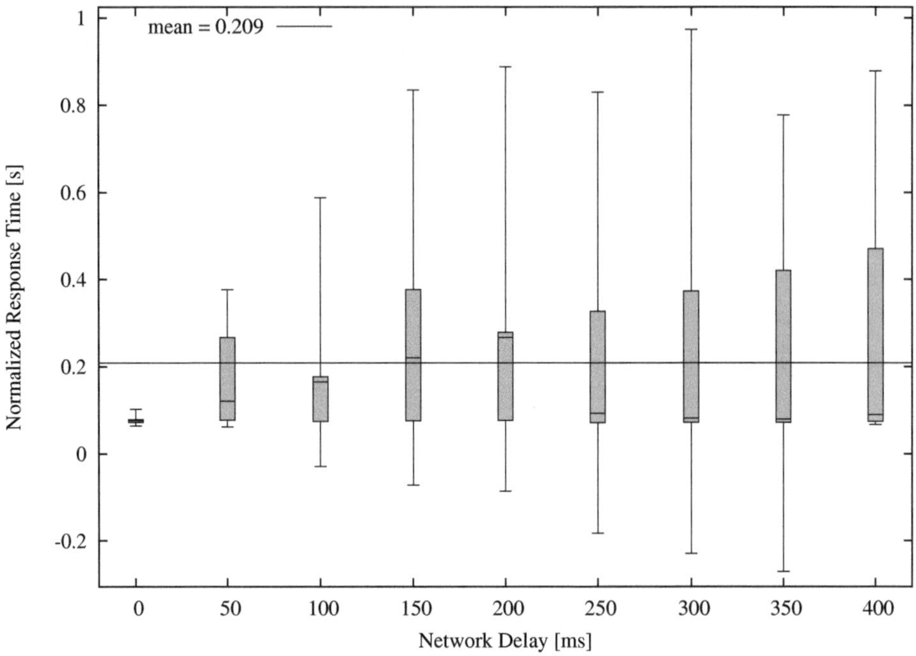

Fig. 4.11: Development of query times of the aggregation.

The execution plan for this second example query, as shown in Figure 4.12, works very similar to that of the first one. Again, the distributed grouping is very efficient. The subsets are grouped as specified in the query. They are then again grouped by gene, and the separate counts added to form the result.

Given that the final result set is still larger than the 11,417 rows sent via the network, this is still the minimum number of rows that has to be transmitted.

4.5.3 *Join*

The join operation shows also very similar results compared to the aggregation and predicate scan benchmarks. It exhibits an increasing spread of results and a shorter execution time as depicted in Figure 4.13. The small boxes indicate that a large portion of the measurements concentrates around a very small area.

In the query plan shown in Figure 4.14, we see that the first part of the query is executed analogically to the simple scan operation. The most important difference is the number of rows sent across the network in between, because the selection returns more data than the first example. The combined intermediary result is then joined together with a local table on the `European` server. Even though the actual

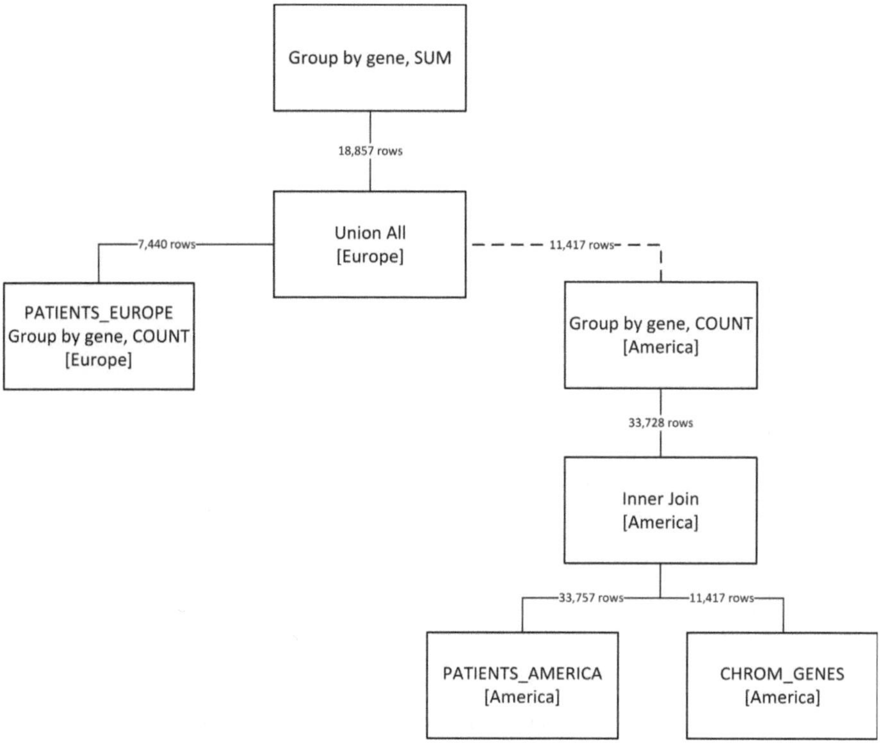

Fig. 4.12: The query execution plan for the aggregation operation. This query returns 13,168 rows. The same optimizations as in the first example are carried out, however more rows have to be sent across the network.

query time is slightly slower than that of the aggregation operation, the execution is faster for higher delays, as less data is transmitted.

It becomes apparent that the IMDB is able to minimize the amount of data sent over the network in this distributed environment, even for more complex queries using aggregation, like the second example. Even though the data is not partitioned using the attribute relevant for aggregation, the IMDB first performs the aggregation in a distributed manner, before merging together the much smaller intermediate results. Furthermore, the execution plans show that as many operations as possible were parallelized, while only the minimum number of rows were sent across the network.

The varying number of packets sent in the second and third benchmark also explains the high standard deviations in the measurements of those operations in comparison to the scan operation. Thus, they do not reflect the benchmark quality in this case, but instead manifest the pattern that is visible in the graphs.

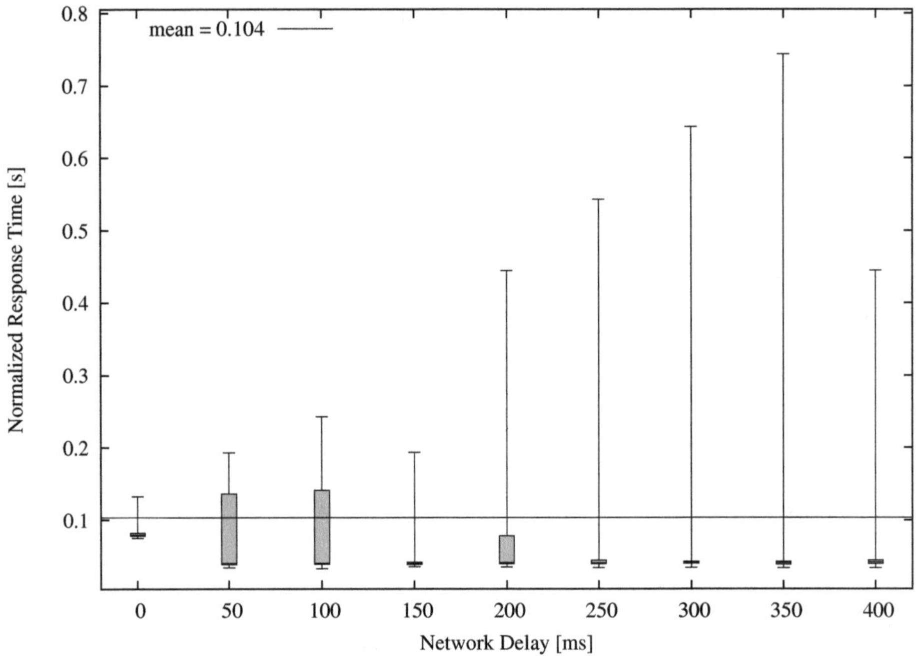

Fig. 4.13: Development of query times of the join.

4.6 Conclusion and Outlook

I demonstrated how apps can work independently from the storage format and location of the data they access. As apparent in programs like the SHARPn project, there is no way around normalizing data and defining standards. However, my approach has the following differences.

- **Runtime normalization:** Instead of prematurely converting data to a standard format, transformation is done only when needed.
- **Application-specific standards:** While I do not try to define and establish any global standards, applications still depend on having stable interfaces they can access. They are allowed to define their own standards, specifying only the parts of the data they need to access.

I created a standardized format for apps to specify database objects and the data they require. Additionally, I introduced a tool that can automatically generate SQL views based on the input format specification of the app. In doing so, I benefit from the capabilities of IMDBs. They make it possible to bring the apps to the data by transforming the latter on-the-fly and connecting databases from all over the world. The specialized query execution plans, which allow working efficiently with

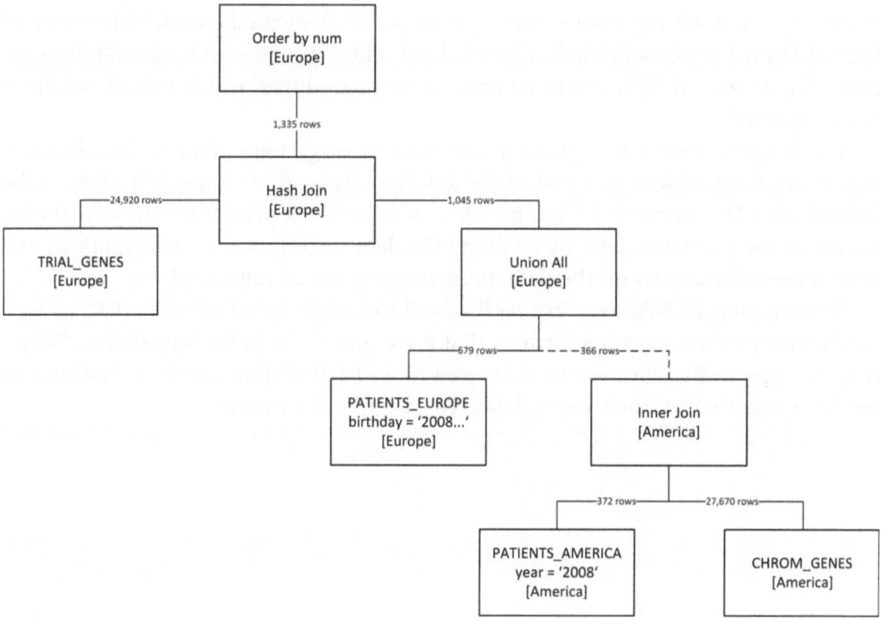

Fig. 4.14: The query execution plan for the join operation. This query returns 1,335 rows.

distributed data, give institutions the chance to collaborate using their heterogeneous data.

Of course, not every institution has the means to manage a resource-intensive IMDB locally. This might be the biggest benefit of centralized services like the BaseSpace platform: individuals willing to share their medical data only have to take care of transforming the data and benefit from the resources in the cloud.

The distribution of tools would be improved through an app store, where they could be uploaded and obtained. This centralized marketplace would benefit from the ability to search apps by the format of data they consume, opening up the possibility of applications interacting with each other.

Enabling apps to work together would then be the next step. Even though applications can specify what data they need, the platform becomes more powerful when apps are able to interact with each other.

For further simplification, the application platform itself could be moved into the IMDB. That way, the app interface is completely decoupled from the backend, consisting of several database objects like tables and procedures. The data would be accessed through a simple web service on top of the database. Some IMDBs like the one used for my benchmarks already offer such a feature.

As a last step, research is needed on how UI for the generation of views for more complex transformations could look like. These UIs need to support complicated mappings of data as well as the combination of multiple data sources. Databases excel e.g. in combining entities through joins, but dedicated engines like those of the SHARPn Project will likely exceed them when it comes to terminology mapping. The power of SQL functions and stored procedures in this regard will have to be explored.

The biggest benefit of IMDBs in the context of genome data is that distribution is not the problem, but part of the solution. Instead of trying to gather all the information from remote locations, we can actually benefit from the distributed nature of the persisted data. By enabling the data store to run transformations and to execute calculations on the data, performance can be improved.

In summary, IMDBs are very well suited to handle large genome data as well as the many heterogeneous formats that are common in today's medicine. Bringing the apps to the data works very well in a distributed network of institutions working together on their mutual data to gain new knowledge.

4.7 References

[83] Bayat A (2002) Science, Medicine, and the Future: Bioinformatics. British Medical Journal 324(7344):1018

[84] Catley C, Frize M (2002) Design of a Health Care Architecture for Medical Data Interoperability and Application Integration. In: Proceedings of the 2nd Joint EMBS-BMES Conference, IEEE, vol 3, pp 1952–1953

[85] Ceri S, Negri M, Pelagatti G (1982) Horizontal Data Partitioning in Database Design. In: Proceedings of the International Conference on Management of Data, ACM, New York, NY, USA, pp 128–136

[86] Chute C et al. (2011) The SHARPn Project on Secondary Use of Electronic Medical Record Data: Progress, Plans, and Possibilities. In: Proceedings of the American Medical Informatics Association Annual Symposium, pp 248–256

[87] Clayton P et al. (1988) Sharing Medical Knowledge for Automated Decision-making. In: Proceedings of the Annual Symposium on Computer Application in Medical Care, pp 591–594

[88] Clayton P et al. (1989) Issues and Structures for Sharing Medical Knowledge Among Decision-making Systems: The 1989 Arden Homestead Retreat. In: Proceedings of the Annual Symposium on Computer Application in Medical Care, pp 116–121

[89] Cock PJ et al. (2010) The Sanger FASTQ File Format for Sequences with Quality Scores, and the Solexa/Illumina FASTQ Variants. Nucleic Acids Research 38(6):1767–1771

[90] Eysenbach G (2008) Medicine 2.0: Social Networking, Collaboration, Participation, Apomediation, and Openness. Journal of Medical Internet Research 10(3)

[91] Giuse D, Kuhn K (2003) Health Information Systems Challenges: The Heidelberg Conference and The Future. International Journal of Medical Informatics 69(2):105–114

[92] Giustini D (2006) How Web 2.0 is Changing Medicine. British Medical Journal 333(7582):1283–1284

[93] Gornik D (2002) UML Data Modeling Profile. Rational Corp, Whitepaper TP162 5(02)

[94] Greenman C et al. (2007) Patterns of Somatic Mutation in Human Cancer Genomes. Nature 446(7132):153–158

[95] Gupta A, Mumick I (1995) Maintenance of Materialized Views: Problems, Techniques, and Applications. Data Engineering Bulletin 18(2):3–18

[96] Hubert B (2001) UNIX Man Pages: tc. http://unixhelp.ed.ac.uk/CGI/man-cgi?tc. Accessed Sep 23, 2013

[97] Illumina Inc. (2013) BaseSpace User Guide. http://supportres.illumina.com/documents/documentation/software_documentation/basespace/basespace_userguide_15044182a.pdf. Accessed Sep 23, 2013

[98] Knöpfel A, Gröne B, Tabeling P (2005) Fundamental Modeling Concepts. Wiley, West Sussex UK

[99] Kuo M, Kushniruk A, Borycki E (2009) An RDF-based Mediator for Health Data Interoperability. Studies in Health Technology and Informatics pp 399–403

[100] McCain KW (1991) Communication, Competition, and Secrecy: The Production and Dissemination of Research-related Information in Genetics. Science, Technology & Human Values 16(4):491–516

[101] Molidor R et al. (2003) New Trends in Bioinformatics: From Genome Sequence to Personalized Medicine. Experimental Gerontology 38:1031 – 1036

[102] Parker C et al. (2004) Detailed Clinical Models for Sharable, Executable Guidelines. Studies in Health Technology and Informatics 107:145–148

[103] Plattner H (2013) A Course in In-Memory Data Management: The Inner Mechanics of In-Memory Databases. Springer

[104] Raden N (2013) On the Persistence of Memory – In Database Systems. http://www.teradata.com/white-papers/on-the-persistence-of-memory-in-database-systems. Accessed Sep 23, 2013

[105] Rea S et al. (2012) Building a Robust, Scalable and Standards-driven Infrastructure for Secondary Use of EHR Data: The SHARPn Project. Journal of Biomedical Informatics 45(4):763–771

[106] Sevilla M (2011) OLAP Databases are being Killed by In-memory Solutions. http://www.capgemini.com/blog/capping-it-off/2011/09/olap-databases-are-being-killed-by-in-memory-solutions. Accessed Sep 23, 2013

[107] Wandelt S et al. (2012) Data Management Challenges in Next Generation Sequencing. Datenbank-Spektrum 12(3):161–171

[108] Wellcome Trust Sanger Institute (2013) How can Mutations Cause Disease? http://www.yourgenome.org/dgg/general/var/var_3.shtml. Accessed Sep 23, 2013

Chapter 5
Billing Processes in Personalized Medicine

Joseph Bethge

Abstract Nowadays, intellectual property such as medical research data is strongly associated with the issue of wanting to protect and share such data for the benefit of research, which prevents disclosure of knowledge. For example, pharmaceutical companies avoid publishing certain parts of their drug discovery research to protect themselves from competitors. As of this writing there are only a limited number of billing systems or market places in the rapidly evolving field of personalized medicine available, where intellectual property, e.g. genome sequencing data, can be traded in exchange for a payment. To contribute, I propose a billing system that can be integrated in an exchange platform for medical data that provides the possibility of receiving payments in exchange for sharing data. Additionally, the system offers the possibility to set fees for specific user groups, which in turn allows to define higher charges for competitors, or let collaborators access certain data for free. In my opinion, exchanging data in this way would benefit both trade partners. As the intellectual property was already gathered, the organizations providing data can use it to collect extra funds. Other institutes buying data can profit by having a larger pool of research data available.

5.1 Introduction

In the year 2000, the vast majority of the human genome had been sequenced in the course of the human genome project [115]. Although it took more than ten years to develop the first draft of the human genome, technology has rapidly evolved ever since, and today it is possible to sequence the whole genome of a human being within 30 hours [122].

Another factor, which drastically changed is the cost for sequencing a whole genome. While the cost was as high as 100 million US dollars in 2001, it decreased to less than 10,000 US dollars in 2011 [138]. It is expected that this trend continues, due to improving technology. Some scientists state that in the year 2020, everyone's genome might be analyzed [128].

This progress enables further scientific research regarding the human genome. Nowadays, instead of being treated on the basis of symptoms, more and more patients get treated based on the mutations in their genome, if the origin of a disease can be found therein. This idea, to treat each patient according to the individual conditions and the cause of a disease, and not based on how other patients with similar symptoms were treated before, is one basic concept behind personalized medicine [124].

There are many institutes all across the world, which gather genome data, for example, to discover why one treatment for a genetic disease helps one patient, but shows no or less effect on another. As such research is still cost-intensive and everyone wants to protect their intellectual property, some organizations avoid to publish the collected genome data for free.

However, what prevents you from offering access to your intellectual property for a fair price? A price you can set freely and change, whenever you think it is too high or too low. That is why I explored how a billing system for a platform, where you can exchange and trade data relevant for genome research, could be implemented. In my opinion, such a billing system would encourage the exchange of important medical research data, and benefit both sides of such an exchange. On the one hand, the owner of the data can collect extra funds, for data, which was gathered anyway. On the other hand, researchers that need additional data, for example, for a control group, can obtain it via such a system.

To allow protection against competitors, different fees can be set in the billing system for certain user groups, and it is also possible to not offer the data at all for specific groups.

Though it is difficult to set a price for an abstract item, such as genome sequencing data, a suggestion feature, which helps users to find a price for an item, could present minimum, average, and maximum prices of similar items to the user. Moreover, if the proposed billing system is part of a platform that provides genome analysis tools in addition to the possibility of exchanging data, you can use the same billing system to collect small usage fees for the analysis tools provided.

The remainder of the work is structured as follows: In Section 5.2, I explain existing billing models and how usage fees can be calculated. In Sections 5.3 and 5.4, I present the requirements and technology needed for my prototype and how my research prototype is implemented. In Sections 5.5 and 5.6, I show and evaluate the benchmark tests I used to prove that my system can be used for several years and still allows to trade items in less than two seconds, and end with a conclusion in Section 5.7.

5.2 Related Work

I present existing billing models, as well as one important example, which provides an exchange platform for genome sequencing data and algorithms to process the data, in the following chapter. Then I will show common ways to calculate usage fees and which factors are used for them.

Existing Billing Models

	Usage-based	Subscription-based
Amount to pay	Based on the usage rate	Fixed
Grants access for	Each unit paid	A certain amount of time
Examples	Pay-per-view, pay-per-click, pay-per-action	Yearly subscription to an online journal

Table 5.1: Comparison of billing models.

Nowadays, there are three billing models, which are generally accepted to be the main models for recurring payments [117]. The usage-based model, the subscription model, which are briefly presented and compared in Table 5.1, and a hybrid model, which combines both models.

In the usage-based model the usage rate for a service or a good is measured in a unit, e.g. in terms of scientific data, which can be data size, amount of files, or the time of use. Examples for these methods are pay-per-click, where clicks on advertisements cost a certain amount, pay-per-action, e.g. you pay for each advertisement that leads a user to registering on your website, or pay-per-view, which gives you access to a file for 24 hours. These methods are applied in the fields of online advertising and platforms for scientific medical publications [112, 127, 137].

In contrast to this, payment methods, which belong to the subscription model, cost a fixed price for a certain amount of time, therefore the price is independent of the actual usage rate. For instance, such models are used for subscriptions to medical online journals and online games [112, 129, Chap. 4.7].

The third possibility combines both usage-based and subscription-based models. For example, a usage-based model for network and Internet access was introduced at Cornell University [136, Sect. 3.1]. The subscription-based model covers the basic usage up to a certain threshold, and any additional usage beyond it, incurs usage-based charges.

Illumina's BaseSpace is one particular example, which provides a cloud platform with an integrated billing system [121]. It allows uploading genome-sequencing data into the cloud for automatic analysis and storage. Furthermore, you can use and develop analysis tools in the cloud platform, and it is possible to share genome sequencing data and these tools with other users.

Although BaseSpace has an existing billing system, it can only be used for apps and not for genome sequencing data. The billing system itself uses a currency, the so-called iCredits, which can be purchased for real money and "may only be used to purchase services, products, or other features available in BaseSpace", but have no monetary value as stated in the terms of use [121].

In contrast to this, it is also possible to use a virtual currency, which is decoupled from real currencies, for example using an approach of a self-recharging virtual currency [123].

Usage Fees for Computer Infrastructures

Developers can choose to run their systems on different computer infrastructures available on the Internet. A popular example with a usage-based billing system for this use case is the Amazon Elastic Compute Cloud (EC2) [135]. Prices for such computing clouds are often dependent on the region, the operating system, and the following resources: CPU power, memory size, storage volume, or data transfer.

The cost can be calculated with a usage-based model, e.g. by the time the resources are used, but there are also hybrid models, for example you can additionally pay a specific fee for one year in a subscription model [116]. Although you still have to pay for the time you use the service you get a discount compared to the model without subscription [135].

5.3 Requirements Engineering

What are the functional requirements you need to know before you can design and implement a billing system? First of all, you need to know for which items you need a payment model. The second question is, how to define the value of an item and how this can be done. You also have to consider any additional functions you require. These questions are answered in the following.

5.3.1 Entity Definition

The following entity definitions shall clarify what data is billed by the proposed billing system. Therefore, I define the structure of that data. A summary of the different entities is given in Table 5.2. The first entity class concerns the elementary basis of genome research, which is genome-sequencing data itself. Before genome-sequencing data are analyzed, it is usually processed in several steps, and after each

Name	Description
Genome sequencing data	The whole or parts of genome data are provided in specific file formats, such as FASTQ, SAM/BAM, and VCF
Gene and mutation annotation	Information about mutations or variants on certain genes or at certain positions in the genome
Electronic health record	Can contain specific data, such as vital parameters, disease history, and information about relatives
Medical paper	A scientific publication, which contains latest research results

Table 5.2: Overview about the tradeable entities.

step a different file format is used. The three relevant formats are: the FASTQ file format, the Sequence Alignment/Map (SAM) format and the Variant Call Format (VCF) format [114, 126, 119].

The second entity class, gene and mutation annotations, covers all possible information snippets about particular positions or a range of positions in the genome, e.g. what can the phenotype of a certain genotype be or which diseases can be caused by a mutation on a particular gene. The HUGO Gene Nomenclature Committee (HGNC) is responsible for approving a gene name and a unique symbol for each known human gene [118]. The committee ensures that each gene is only given one approved gene symbol and differentiates it from other aliases. This makes the gene symbols and aliases provided by the HGNC an ideal choice to identify the annotations' genes. There are multiple databases, which can identify single mutations. I chose the dbSNP database from the National Center for Biotechnology Information (NCBI) as one example. The database includes 53,567,980 mutations, 38,072,522 of which are validated as of build 137 on June 26, 2012 [130].

Electronic Health Records (EHRs) and Electronic Medical Records (EMRs) are the third entity class. They can be very valuable when available together with genome data. That is because they enable automated analysis, based on genome data in combination with, for example, environmental or medication data, e.g. for clustering of patient cohorts as described in Chapter 3 or to find proper clinical trials for a patient as described in Chapter 8. The contents of EHR and EMR are basically the same, but the scope is different, an EMR only contains the data needed by one physician, while an EHR can contain data from all physicians and diseases a patient had in his life [113]. For example, SAP EMR System for mobile devices contains the following data: lab results, doctor's appointments, attending physicians, address, insurance information, information about relatives, and vital parameters, such as temperature, blood pressure, or pulse data [133].

Medical papers are the last entity class. They are a common way to exchange medical research findings. Therefore, they are exchangeable as files, such as documents stored in PDF format.

5.3.2 *Free Price System*

First of all, the currency I chose for my billing system, is a virtual currency called Gene Points (GP), because I want to use an independent currency for my prototype. Once the billing system is used for a real application, it is still possible to set exchange rates to other currencies later. To get started, every user receives a specific amount of GP as an initial investment.

With the free price system, I refer to the right and the responsibility to freely set prices for items you want to sell on a market. This allows finding a price, even for items, which are difficult to determine a price for, by leaving this process up to the people trading these items. To help users find an appropriate price for a new item, the billing system provides minimum, average, and maximum prices for the type of item.

One problem is, what value does the first item of a kind have? For example, if you are the first one to upload genome-sequencing data, at which price should you offer it? For the first item you would have no comparison available and therefore you and all potential buyers do not know whether a price you choose is high or low. As the supply of data is practically zero, you would probably like to choose a very high amount of GP. The problem is you do not know what a very high amount of GP is.

However, there is a possible solution for it, which is a minimum and a maximum limit for each item type. Because these limits should be estimated from experts in the corresponding field, I assumed some value ranges for my research prototype, which can be seen in Table 5.3.

The range of these limits depends on the expected value differences between the most and the least valuable item of one type. Of course, the relation between these intervals also reflects the relation between the values of the corresponding item types.

For example, I expect that medical papers with important research findings can be much more valuable than a single gene annotation, therefore I presumed a higher minimum limit for medical papers than maximum limit for gene and mutation annotations. Another example is, the value difference between the whole sequenced genome from a person with a unique disease and a small sequence snippet without any mutations, is probably much larger than between publishing medical papers, which all require scientific work and research. Therefore, I presumed that the interval of genome sequencing data spans three orders of magnitude, compared to the one order of magnitude for medical papers.

Another possibility is, to consider these limits only as soft limits, which only advise against exceeding them. For example, this would allow items of unusual high value to still be offered at their value. The minimum limit should be a soft limit in any case, as it allow items to be provided free of charge.

Furthermore, I conducted user interviews, which resulted in the following additional requirements. There should be a possibility to define user groups and item groups. Why can this be useful? On the one hand, you can define different prices for different user groups, for example, remove the price for other institutes, which

Name	Minimum	Maximum
Genome sequencing data	10	10,000
Gene and mutation annotation	1	100
Electronic health record	50	500
Medical paper	1,000	10,000

Table 5.3: Example price ranges assumed for my research prototype.

you have a research agreement with, and share your data at no charge, while still offering it to others for the original price. On the other hand, in itself, without corresponding medical data, genome sequencing data often is not particularly interesting. In contrast to this, genome sequencing data together with additional information, can become much more valuable for certain research questions. Therefore, you should also be able to sell such combined data as one unit.

5.3.3 Calculation of Usage Fees

Usage fees can and should be charged whenever a user uses substantial parts of the system performance. For example, trading items via my billing system does not need a lot of system resources, but if a user wants to use the system for processing of whole genome sequencing data, e.g. alignment and variant calling of a large FASTQ file, the whole system may be blocked for several minutes [134].

To provide fair usage fees for such cases, you need to know the cost to use and maintain the system for any given time, and how much time a task exactly needs. This allows to compute the actual cost incurred, by multiplying the time with the cost per time. Therefore, the system provider adds up the running costs, e.g. for electricity, Internet traffic, software licenses, and the costs for the server nodes.

For instance, a user starts the processing of genome sequencing data at a certain time. After it is finished, the billing system analyzes the statistics of the whole task, e.g. which part took how long, then sum up all the time used and charge the user with the amount of GP accordingly.

Note that the system can still stay decoupled from real currencies. A possible approach for this is named in Section 5.2, which can be applied to this use case in a similar way. You calculate the amount of GP users paid as usage fees for specific time intervals. This amount of currency left the market system and can be injected by distributing an equal part of that amount to all users. It is possible to wait until a certain time period has passed and therefore reduce the amount of GP users can spend to use the system.

For example, take only two active users, with a sufficient amount of GP, to pay the usage fees for twelve hours of using the whole capacity of the system. Further, assume a time interval of one day, without any waiting period. Both users can spend all their GP and the system will be under full load for 24 hours a day. The next day both gain the same amount of GP back, as they are equally distributed. In contrast to this, if you redistribute the GP only after one week, these two users can only use the system for twelve hours a week. That is how such a delay can be used to control the load of the system.

5.3.4 In-memory Database Technology

In this section, I explain why I decided to use in-memory database technology for my billing system, and in particular how it is advantageous for my use case.

To show users all the transactions they paid for or received from other users, you have to keep a log of all transfers between the different users. This also makes it possible to compute the actual balance of a user, without the need to save it in a different database table, as you can calculate it by subtracting the amount of GP a user paid from the amount of GP a user gained.

Now assume the billing system is used over a long period of time. This means the transaction log table will get larger and larger, and the time to add and subtract all the transfer values is getting higher eventually. But, of course you still want to maintain short answer times for users and fast processing times in your database for all database requests.

Stored Procedures

Moving application logic directly to the data stored within the database is beneficial for data processing, e.g. with the help of stored procedures [131]. But why is this desirable? First of all, stored procedures centralize functionality in the database and can therefore be reused by all applications, which need these functions. As a result, the amount of application code is reduced and changes to these functions are simplified.

A second advantage, which is mentioned, is the reduction of network traffic and network latency. More complex tasks often need multiple SQL queries, where the output of one query gets processed in an application or a function and is used as an input of other queries.

Another advantage is the atomicity of stored procedures. That means, if something during the execution of a stored procedure fails, the statements executed so far are rolled back, which always promises a consistent database state. For example, if you transfer money from one account to another, you want to subtract the money from the sender and add it to the recipient. What you do not want, is that the recipient retrieves the money, but the sender does not lose any or vice versa. Thus, the use of proper isolation levels ensures the consistency of database transactions.

Column-oriented Storage

In-memory databases offer row- and column-oriented storage [110]. As join and aggregation queries operate on a subset of columns instead, the column-oriented storage avoids accessing data not required to compute the output of such queries.

For example, the transaction log table consists of the following columns: `user`, `item`, `type` of item, `recipient`, `time`, `price` and the `validity period` of the item. To calculate the balance of a user you only need three out of these seven columns, the user, the price, and the recipient. In contrast to this, if you want to analyze the maximum, average, or the minimum price for a type of items, you only need the type of the item and the price.

Lightweight Compression

Another feature of the incorporated in-memory databases is lightweight compression, which has two main goals. On the one hand, it "is still expensive to process [...] huge data sets entirely in main memory", which creates the need for compression [131]. On the other hand, lightweight compression techniques have the potential for improved query performance, as less data has to be transported between main memory and CPU. For example, compression techniques useful for different types of column contents are discussed by Abadi et al. [109]. These techniques can be applied to the columns separately, and are able to reduce the table size, which is especially important for the tables that are expected to need the most space, for example the table storing all transactions.

Partitioning and Parallel Computing

To accelerate the analysis of the very large transaction log database table, the advantages of partitioning and parallel computing can be used. In in-memory databases the concepts of vertical and horizontal partitioning can be used to split the contents of a database table. Tables with the aforementioned column-oriented storage, partitioning can be used for large tables to split them horizontally and distribute them across different nodes.

This allows aggregates to be computed in parallel for each of these distributed partitions, e.g. each CPU core can search one partition. This reduces the total time needed for a query, especially if you take advantage of the different partition possibilities, such as position or value range partitioning, hashing, or round robin [131].

For example, if you have the transaction log of a billing system, and all queries only need the entries affecting a certain user, you can use hash partitioning on the corresponding user column. If you have a single user column, this distributes all transactions affecting one user in the same partition. Thus, only a part of the whole table has to be analyzed.

5.4 Application Example

In this chapter, I will explain the architecture and implementation details of my prototype, and the integration into the prototype of the High-performance In-memory Genome (HIG) project as introduced in Section 1.5. Therefore, it is possible to make use of the user authentication available in the HIG system, and more importantly, the possibility to upload and analyze genome-sequencing data.

The HIG system architecture is modeled as Fundamental Modeling Concepts (FMC) block diagram in Figure 5.1 and it consists of the following layers: application, platform and data layer [125, 134].

For the billing system, I added new functionality in all of the aforementioned layers. I extended the data and platform layer by specific database content and stored procedures as described in Section 5.4.1 and Section 5.4.2 respectively. Furthermore, I implemented the graphical user interface in the application layer as described in Section 5.4.3 and the corresponding billing protocol as described in Section 5.4.4.

5.4.1 Database Schema

In the following section, I explain the relationships between the objects I store in the in-memory database. In Figure 5.2 the most interesting part, needed to offer flexible prices with support for item groups and user groups as described in Section 5.3.2, is shown.

At first, there is the user of the system. A user, for example, can be a researcher or an institution, and is represented by a unique user ID. A user can upload multiple items into the system, which he will then possess, while an item can have one owner at most. An item type and an item number identify an item. As of this writing, the HIG platforms allows data upload without being logged in, as well. Therefore, an item can exist without an owner, in this case the item is not billed by the billing component, and is automatically available for free.

Both users and items can be arranged in groups. A user can belong to multiple groups, but does not have to be part of one, while a group can have multiple users, and possibly none, for example if the only member of a group leaves the system. This applies for items and item groups in the same way, although you cannot create an item group without any item in it.

A group of items can now be offered for multiple user groups with a different price and duration for each user group. Therefore, offers can be created, which always belong to exactly one user group and one item group, and set the price this user group has to pay for a certain duration of access to data or a service. Although the duration field implies a subscription model, the duration can also be very short or unlimited, which in turn can make the payment option similar to the usage-based models explained in Section 5.2, such as pay-per-view.

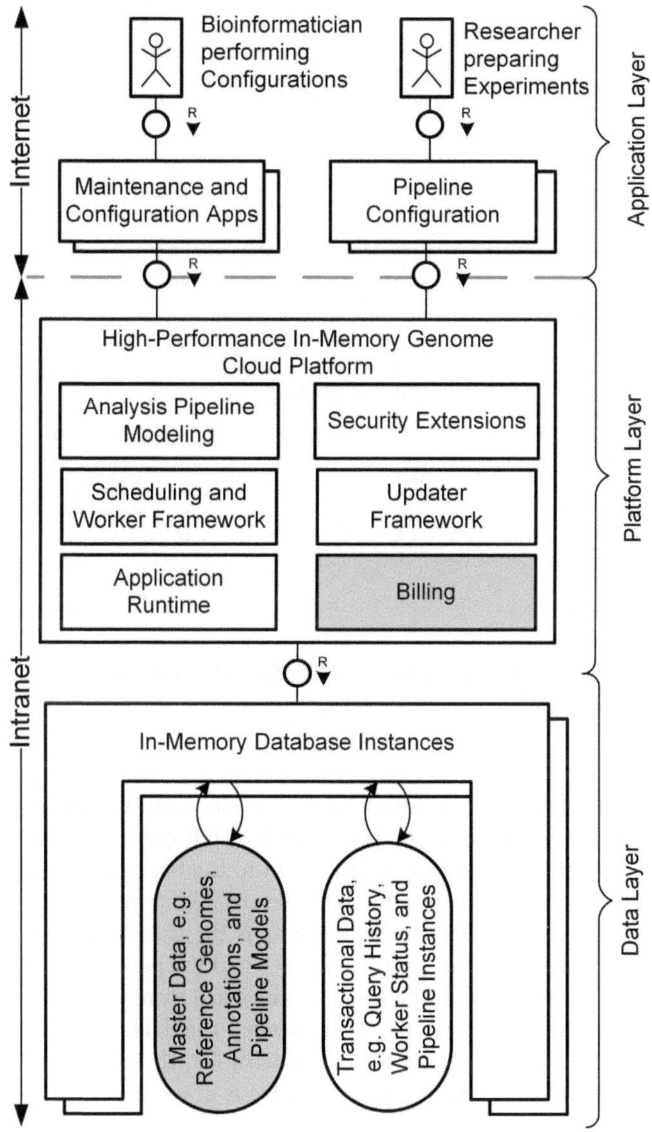

Fig. 5.1: Integration of billing functionality in HIG system modeled as FMC block diagram.

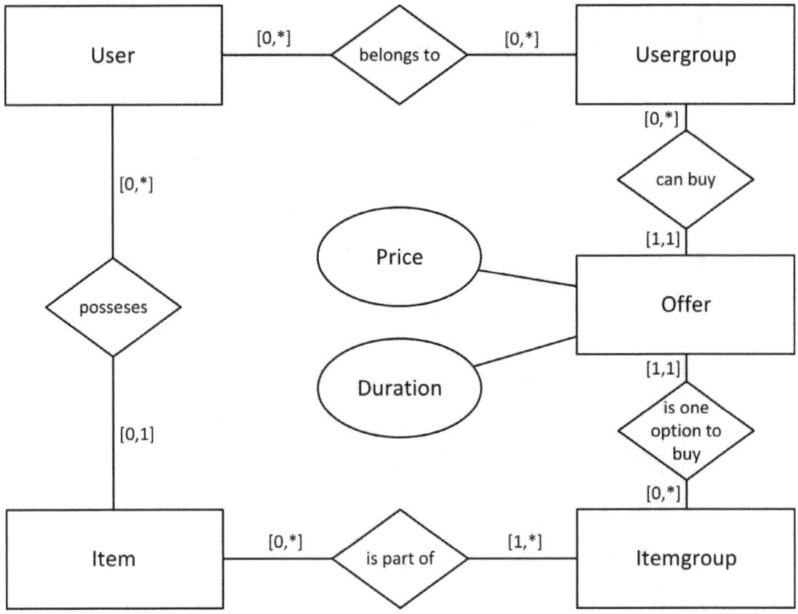

Fig. 5.2: Involved entities for billing to offer flexible prices for different users and groups of items modeled as entity relationship diagram.

However, there are some additional database elements, which are not included in the entity relationship diagram. For example, I left out many attributes, which are only important for displaying information to the user, such as group descriptions, or the entity, which represents the different item types.

Another table not included in the diagram, is the transaction log table, which stores all information relevant for the balances and the purchased access rights of users. For each transaction, the following values are stored in this table: the user ID of the sender, the item type, the item number, the data and time at which the transaction was created, the transferred value in GP, the user ID of the receiver, and the duration of how long access to the item is granted.

5.4.2 Database Functionality

My aim is to implement the whole application logic of the billing component as stored procedures within the database. Therefore, I use stored procedures for all functions needed from the database, which require more than just one select statement. For example, the procedure shown in Listing 5.2, executes multiple SQL statements internally as an atomic transaction. In contrast to this, if a function

```
SELECT DESCRIPTION, ID FROM BILLING.GROUPS
WHERE OWNER_ID = #{current_user_id} OR OWNER_ID = -1
```

Listing 5.1: The statement used to retrieve user groups. Single SQL SELECT statements are executed directly.

```
CALL BILLING.BOOK (#{user}, #{type}, #{item}, #{exp_price}, #{
    exp_duration}, ?)
```

Listing 5.2: The procedure call to buy an item. All functions that perform multiple statements, e.g. INSERT statements, are encapsulated within a stored procedure.

retrieves all user groups a user can access, e.g. they were created by the user or are public, a single select statement, which can be seen in Listing 5.1, is sufficient.

I use the atomicity of stored procedures to guarantee that the functions executed in the database never return an inconsistent state of the database. Another issue I considered is locking the database table and rows. This is needed to ensure that a user with 50 GP cannot quickly buy two items worth 50 GP each, and end up with -50 GP. For example, if you lock the transaction log table, before calculating the balance and inserting the transaction, the called procedure for the second item, waits with calculating the balance, until the first procedure releases the lock. Therefore, it is correctly recognized that the user has 0 GP, and the second procedure call will return, stating that the balance is insufficient.

5.4.3 User Interface

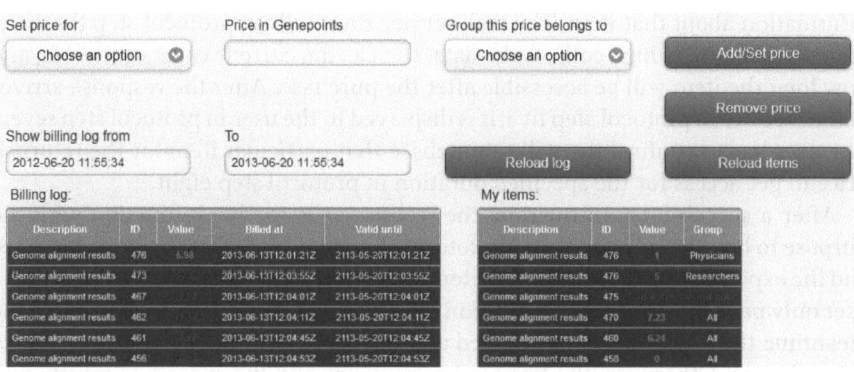

Fig. 5.3: Screenshot of the user interface for management of item prices.

First of all, a user who visits the website of the HIG system, can see prices, shown for all available genome sequencing data items, which are displayed on the main page. Access to items, which are not free, have to be purchased with my billing system. The process of buying an item is further explained below as an example of how the different components communicate.

There is also a new tab visible, which is called “Billing”. The purpose of this tab is to review all transactions from the past, which affect you, and to manage the prices of items you possess. If a user clicks it, a website similar to the one seen in Figure 5.3 is displayed.

This website shows the transactions of the last year, but the time range of the displayed items can be changed, e.g. to narrow down the transactions of specific time periods. Furthermore, users can click on one of their items, which will automatically select this item, the price and the group, and select or insert these values in the corresponding entry fields and select menus, which are labeled with “Set price for”, “Price in Gene points” and “Group this price belongs to”. A user can now change the price or the group and click the button “Add/Set price” to either update the price or add it as a new option, if no price was specified for the selected group yet. If a user clicks “Remove price”, it will remove this particular offer for the selected item and the group.

5.4.4 Billing Process

How can users buy genome-sequencing data? This can be done in the same way as data has been accessed without my billing component, only with one added interaction with the user. For example, before I integrated my billing system a user could click on a genome sequencing data item on the main page of the HIG platform, which would direct him to the shown item. However, if a user clicks on such an item now, the process illustrated in Figure 5.4 starts.

In protocol step two, the website calls the function in the web service to get the information about that item. The web service then calls in protocol step three the stored procedure, which collects the data, such as the current value of an item, and how long the item will be accessible after the purchase. After the response arrives at the website in protocol step five, it is displayed to the user in protocol step seven. The user has to confirm the will to purchase that particular item for the returned price to get access for the specified duration in protocol step eight.

After a successful confirmation, the website calls the book function with the purpose to buy the selected item in protocol step nine. Note that the expected price and the expected duration are parameters of this call. I implemented it this way, so a user only pays the price for the duration he confirmed. If the price is changed in the meantime the book request is rejected and the user is informed about the change price change. Otherwise, the transaction succeeds and the success is returned to the website in protocol step 15.

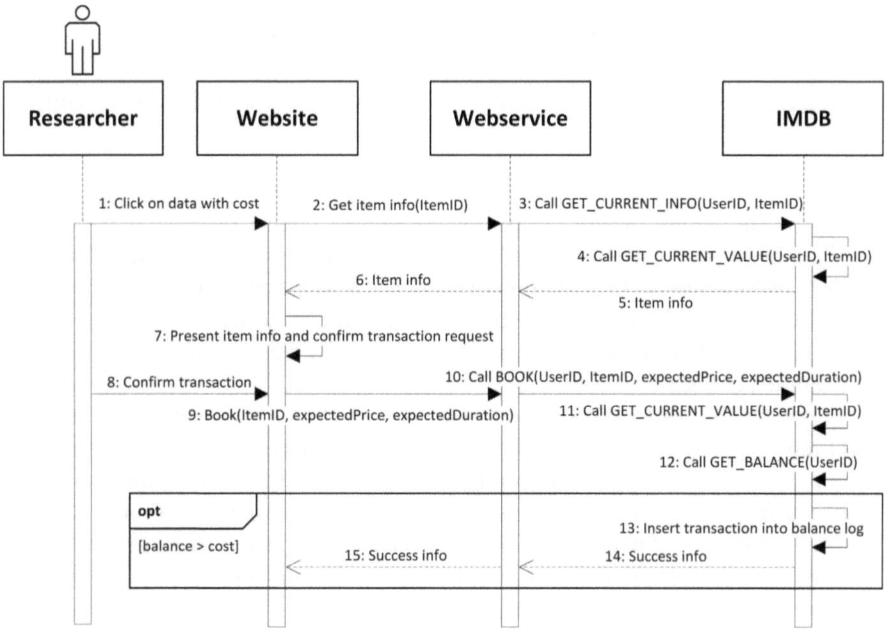

Fig. 5.4: Performed process steps when accessing a priced item modeled as UML sequence diagram.

5.5 Benchmarks

In this chapter, I describe the details of my conducted benchmarks. My motivation is to prove that the system is capable to handle data for at least a ten year period of time while the response time for transaction is less than a two seconds threshold [120]. I also want to verify that data partitioning as introduced in Section 1.4.4 is a possible tuning factor to control the response time.

In order to find a realistic size of data sets for testing purposes, I conducted user interviews with researchers, who deal with genome sequencing data. The identified use case for researchers is to purchase genome-sequencing data with additional details, such as medical data of the individuals, to validate findings or as additional control data. In such a scenario, researchers may want to buy a set of data every few months.

Nowadays, clinical trials, which provide genome-sequencing data, have sequenced samples of a few hundred patients only as described in Section 6.4.1. Thus, I presume that a group of researchers would access some hundred items of genome sequencing data together with the corresponding EHR. For example, I assume a set of about 700 items to be traded each quarter. How many groups of researchers could trade such items via my billing system?

```
CALL BILLING.GET_CURRENT_INFO(752, 1, 751, ?)
CALL BILLING.BOOK(752, 1, 751, 25, -1, ?)
```

Listing 5.3: The two statements, filled with example data, used for the benchmark.

One of the largest organizations associated with cancer research worldwide, is the American Association for Cancer Research (AACR), which currently has more than 34.000 members [111]. Although there are also membership categories, such as a student or an honorary membership, I assume a number of 35,000 users for my system, as all of these members could potentially start researching in the field of genome analysis.

In conclusion, this means, if 35,000 users trade 700 items every three months, after ten years the transaction log contains 980 millions of transactions. Therefore, I vary the size of the log, which contains between 100,000 entries and one billion transactions of four thousand items randomly traded between one thousand users.

For the purpose of benchmarking, I used an IMDB system installed on two identical servers configured as database landscape as detailed in Table 5.4.

Processor	8 x Intel® Xeon® E7-8870 @ 2.40 GHz
Main Memory	64 x 32 GB DDR3 1066 MHz + 64 x 16 GB DDR3 1066 MHz
Operating System	SUSE Linux Enterprise Server 11.2 (Kernel 3.0.13-0.27-default)

Table 5.4: Configuration of benchmark system.

I measure the time in a Python script, with the time function of the equally named module of Python. The documentation states that the precision of this function "is much more precise" than one hundredth of a second, so I presumed the precision is at least 0.01 s [132]. The queries shown in the Listing 5.3 are required to complete a buying transaction for an item with my billing system as depicted in Figure 5.4.

In addition to varying the number of entries in the transaction log, I also measured the needed time for these queries on both a partitioned and a non-partitioned transaction log table. I selected a two-level hash partitioning with eight partitions in total. I used the attribute describing the sender of a transaction as first level criteria and the attribute describing the recipient of the transaction as the second level criteria. The exact SQL statements to create the table with and without partitioning can be found in Listing 5.4 and Listing 5.5.

I choose 1,000 as a sample size, so I can execute the query for all different sample users I created. My goal was to reach a standard deviation below 20 percent.

The results of my benchmark tests for the partitioned and non-partitioned transaction log table, can be seen in Table 5.5 and Table 5.6 respectively.

The differences between the number of entries in the partitions of one table is lower than 30 % for all partitioned tables. For example, in the partitioned table

	Get Value		Book	
#Transactions	Mean [s]	SD [%]	Mean [s]	SD [%]
101,000	0.146	2.31	0.181	6.19
1,001,000	0.147	2.79	0.182	5.94
5,001,000	0.144	2.73	0.184	5.56
10,001,000	0.143	2.70	0.185	5.60
50,001,000	0.145	2.76	0.282	5.14
100,001,000	0.145	2.86	0.335	5.38
506,636,000	0.124	8.48	0.400	9.97
1,000,001,000	0.163	15.05	0.768	7.39

Table 5.5: This table shows the average query time as well as the standard deviation (SD) for a partitioned transaction log table.

	Get Value		Book	
#Transactions	Mean [s]	SD [%]	Mean [s]	SD [%]
101,000	0.137	2.55	0.160	2.71
1,001,000	0.147	2.62	0.166	2.34
5,001,000	0.137	2.48	0.175	2.26
10,001,000	0.139	2.41	0.185	2.22
50,001,000	0.181	2.07	0.317	1.76
100,001,000	0.234	3.47	0.643	2.51
592,319,000	0.738	6.27	3.878	3.58
1,000,001,000	1.203	5.90	6.553	3.51

Table 5.6: This table shows the same values as Table 5.5, but for a non-partitioned transaction log table (SD = Standard deviation).

with 1,000,001,000 entries, the partition with the highest and the lowest number of entries has 137,000,502 and 111,000,010 entries, respectively.

The compression ratio between the size of the Comma-Separated Values (CSV) files used to import the data into the database and the size of the data in-memory is up to 5.9:1. For example, the size of the CSV file for 101,000 entries was 4,985 KB, compared to 858 KB of total memory consumption in the IMDB.

5.6 Evaluation

In this chapter, I discuss how the measured values verify my hypotheses stated in Section 5.5. Then I describe possible influences on my test results.

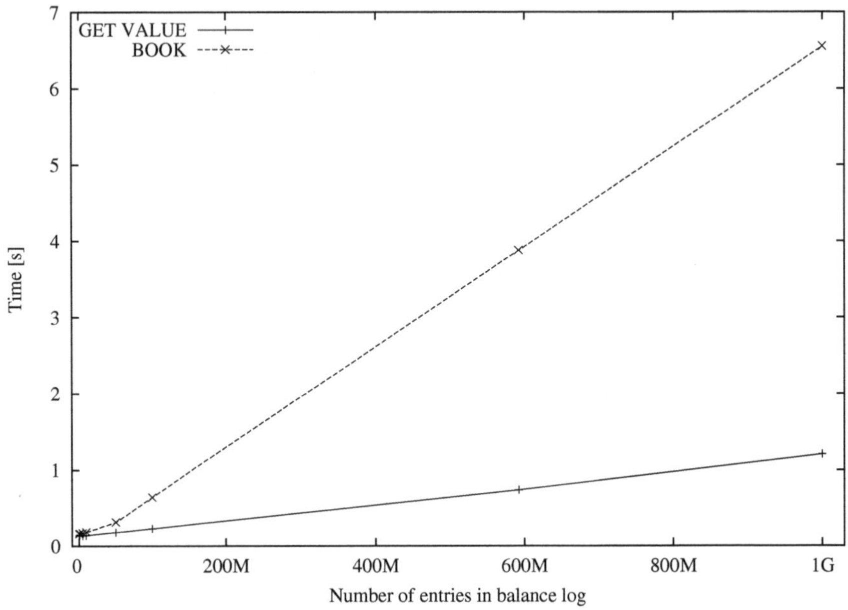

Fig. 5.5: Run time to execute both queries for different transaction log sizes.

5.6.1 *Impact of Transaction Log Size*

As the size of the transaction log increases, the execution time of both queries increases as depicted in Figure 5.5. This correlation seems to be a linear one, which can also be seen in the implementation of these queries.

The procedure to retrieve the current value of a specific item for a user counts the transactions in the balance log, which might already affect this item. If there are any entries, the date and duration of the last access period granted to that user is compared to the time the query was executed at. If an entry is still valid, the user can still access that item, and is not required to pay. In this case, the function returns the amount of time the item is still accessible. However, if there is no such entry, the price and the duration for the item are returned.

This particular procedure is called in both procedures, which are used in the tests. As the time needed for the aggregations on the transaction log table is dependent on its size, the time needed for both procedures increases. In fact, at about one billion entries the book procedures need approximately 6.5 s to buy one item. In my opinion, this waiting time is far too long for a user. But this occurred without partitioning the transaction log table (see Listing 5.5 for the exact create statement).

5.6.2 *Impact of Data Partitioning*

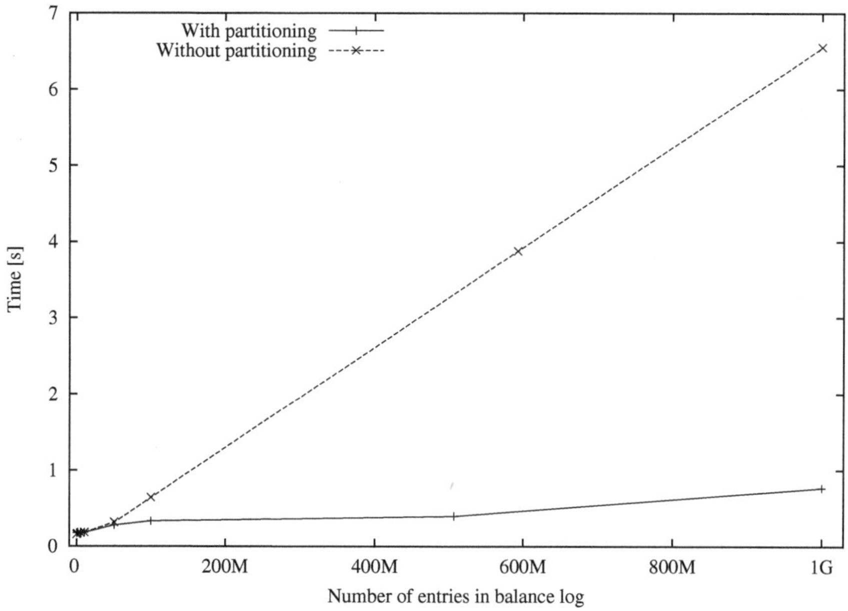

Fig. 5.6: Procedure run time of BOOK.

Now what happens, if we partition the table based on the hash of the values in those columns that are used to select the correct rows? The source code for this can be seen in Listing 5.4. The aggregations on the table could speed up by a factor, which equals to the amount of partitions since the amount of rows, which have to be aggregated are divided by the same factor.

In fact that is exactly, what the query times suggest, as it can be seen in Figure 5.6, or in the comparison of the plain numbers in Table 5.6 and Table 5.5. The book statement with the same one billion entries in the transaction log table, now needs only about 0.768 s in average, which is even less than one eighth of the time need without partitioning. It is also possible to observe from the test results that the time difference between both measured procedures increases. This implies, the book procedure has an additional part, which needs a longer time to be calculated, for a higher amount of entries in the transaction log table.

The book procedure calls the procedure to calculate the balance of the user. This is done by aggregating the debits and the credits of the user out of the transaction log, and subtracting them accordingly. That is why the size of the transaction log table further influences the time needed for the book procedure.

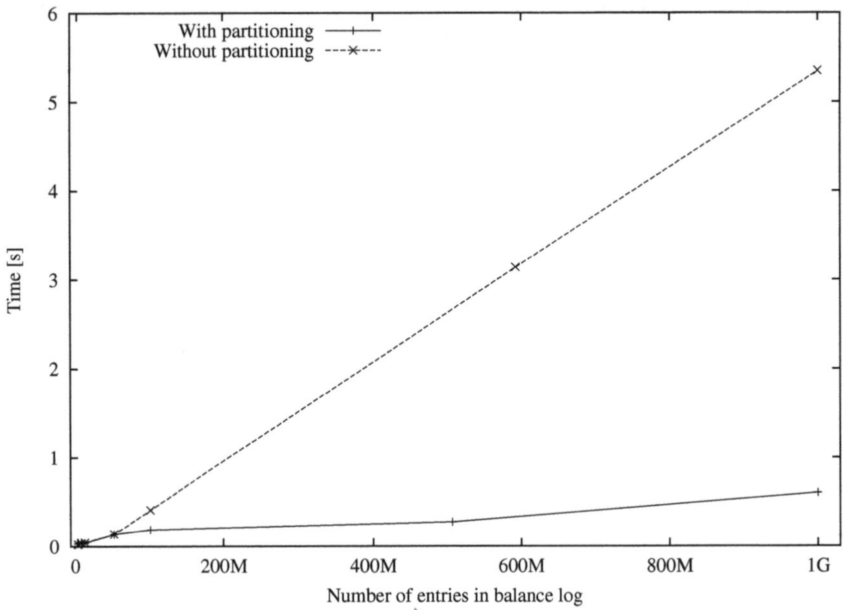

Fig. 5.7: Comparison of procedures BOOK and GET VALUE.

5.7 Conclusion and Outlook

In this work, I proposed a billing system, which allows trading of intellectual property, with the focus on medical research data, concerning the processing and analysis of genome data.

I selected specific types of data relevant for genome analysis research and defined the formats in which they can be exchanged on a platform. According to these data types I investigated and implemented flexible payment models, such as pay-per-use, prepaid, and subscription models. To be able to offer reasonable prices, I use a free price system, where users can freely decide, which payment model they want to use for their data and adopt the price according to the number of accesses and prices stated for similar data.

Furthermore, I showed, how usage fees for computer infrastructures can be applied to processing of genome sequencing data. I presented one way of integrating such a billing system into an in-memory database, and the advantages this integration provides.

Moreover, I implemented a prototype, and integrated the functionality into the existing HIG platform. To prove the feasibility of such a system running for multiple years, I executed benchmark tests on my system. The results proved that the time needed the buy an item remains below the two seconds threshold even for

```
CREATE COLUMN TABLE "BILLING500M"."BALANCE_LOG" ("USER_ID"
    INTEGER CS_INT,
   "ENTITY_ID" INTEGER CS_INT,
   "ENTITY_OBJECT_ID" INTEGER CS_INT,
   "BILLED" LONGDATE CS_LONGDATE NOT NULL ,
   "VALUE" DOUBLE CS_DOUBLE,
   "OWNER_ID" INTEGER CS_INT,
   "DAYS_DURATION" INTEGER CS_INT) WITH PARAMETERS ('
       PARTITION_SPEC' = 'HASH 4 USER_ID; HASH 2 OWNER_ID');
```

Listing 5.4: SQL code for a partitioned transaction log

```
CREATE COLUMN TABLE "BILLING500MNP"."BALANCE_LOG" ("USER_ID"
    INTEGER CS_INT,
   "ENTITY_ID" INTEGER CS_INT,
   "ENTITY_OBJECT_ID" INTEGER CS_INT,
   "BILLED" LONGDATE CS_LONGDATE NOT NULL ,
   "VALUE" DOUBLE CS_DOUBLE,
   "OWNER_ID" INTEGER CS_INT,
   "DAYS_DURATION" INTEGER CS_INT);
```

Listing 5.5: SQL code for an unpartitioned transaction log

one billion items. This is a reasonable expectation in the field of medical research data, after the system has been used for ten years.

Future work could discuss, if such an approach can be used in other fields, and what has to be adapted in that case. This might lead to similar, more flexible, billing systems for intellectual property in fields other than medical research.

5.8 Appendix

This appendix contains selected parts of the source code from my prototype. Listing 5.4 shows the SQL query for creating a partitioned transaction log table. Listing 5.5 shows the same table definition for an unpartitioned transaction log.

5.9 References

[109] Abadi D et al. (2006) Integrating Compression and Execution in Column-oriented Database Systems. In: Proceedings of the International Conference on Management of Data, ACM, New York, NY, USA, pp 671–682
[110] Abadi D et al. (2009) Column-oriented Database Systems. Proceedings of the Very Large Data Bases Endowment 2(2):1664–1665
[111] American Association for Cancer Research (2001) Membership Mailing Lists. http://www.aacr.org/home/membership-/become-a-member.aspx. Accessed Sep 23, 2013
[112] American Association for Cancer Research (2012) Subscriber Help & Services. http://www.aacrjournals.org/site/subscriptions/index.xhtml. Accessed Sep 23, 2013
[113] American Medical Association (2013) Electronic Medical Records and Electronic Health Records. http://www.ama-assn.org/ama/pub/physician-resources/health-information-technology/health-it-basics/emrs-ehrs.page. Accessed Sep 23, 2013
[114] Cock PJ et al. (2010) The Sanger FASTQ File Format for Sequences with Quality Scores, and the Solexa/Illumina FASTQ Variants. Nucleic Acids Research 38(6):1767–1771
[115] Collins FS, McKusick VA (2001) Implications of the Human Genome Project for medical science. The Journal of the American Medical Association 285(5):540–544
[116] Deelman E et al. (2008) The Cost of doing Science on the Cloud: The Montage Example. In: Proceedings of the Conference on Supercomputing, IEEE Press, Piscataway, NJ, USA, pp 50:1–50:12
[117] Dibble, Tom (2012) Finding a Billing Model to Fit Your Startup. http://www.ecommercetimes.com/story/75577.html. Accessed Sep 23, 2013
[118] European Bioinformatics Institute (2009) The HUGO Nomenclature Committee. http://www.genenames.org/about/overview. Accessed Sep 23, 2013
[119] European Bioinformatics Institute (EMBL-EBI) (2011) VCF (Variant Call Format) version 4.0 | 1000 Genomes. http://www.1000genomes.org/wiki/Analysis/Variant19,2013. Accessed Sep 23, 2013
[120] Galitz W (2002) The Essential Guide to User Interface Design: An Introduction to GUI Design Principles and Techniques. John Wiley & Sons, New York, NY, USA
[121] Illumina, Inc. (2013) BaseSpace Credits Purchase and Terms of Use. https://basespace.illumina.com/agreements/store. Accessed Sep 23, 2013
[122] Illumina, Inc. (2013) HiSeq 2500/1500. http://www.illumina.com/systems/hiseq_2500_1500.ilmn. Accessed Sep 23, 2013
[123] Irwin D et al. (2005) Self-recharging Virtual Currency. In: Proceedings of the Workshop on Economics of Peer-to-peer Systems, ACM, New York, NY, USA, pp 93–98
[124] Jain K (2009) Textbook of Personalized Medicine. Springer

[125] Knöpfel A, Gröne B, Tabeling P (2005) Fundamental Modeling Concepts. Wiley, West Sussex UK
[126] Li H et al. (2009) The Sequence Alignment/Map (SAM) Format and SAMtools. Bioinformatics 25:2078–2079
[127] Mahdian M, Tomak K (2007) Pay-per-action Model for Online Advertising. In: Proceedings of the 1st International Workshop on Data Mining and Audience Intelligence for Advertising, ACM, New York, NY, USA, pp 1–6
[128] medGadget (2012) The Human Genome, Sequenced: Predicting the Future of DNA Tech. http://www.theatlantic.com/health/archive/2012/03/genome-sequenced/254068/. Accessed Sep 23, 2013
[129] Nae V et al. (2011) A New Business Model for Massively Multiplayer Online Games. In: Proceedings of the 2nd International Conference on Performance Engineering, ACM, New York, NY, USA, pp 271–282
[130] National Center for Biotechnology Information (2012) dbSNP Summary. http://www.ncbi.nlm.nih.gov/SNP/snp_summary.cgi. Accessed Sep 23, 2013
[131] Plattner H (2013) A Course in In-Memory Data Management: The Inner Mechanics of In-Memory Databases. Springer
[132] Python Software Foundation (2013) 26.6. Timeit - Measure Execution Time of Small Code Snippets. http://docs.python.org/2/library/timeit.html. Accessed Sep 23, 2013
[133] SAP AG (2011) SAP Community Network Wiki - Healthcare - Electronic Medical Record (EMR). http://wiki.sdn.sap.com/wiki/display/HC/Electronic+Medical+Record+19,2013. Accessed Sep 23, 2013
[134] Schapranow MP, Plattner H, Meinel C (2013) Applied In-Memory Technology for High-Throughput Genome Data Processing and Real-time Analysis. In: Proceedings of the XXI Winter Course of the Centro Avanzado Tecnológico de Análisis de Imagen, pp 35–42
[135] Services AW (2013) Amazon Elastic Compute Cloud (Amazon EC2), Cloud Computing Servers. http://aws.amazon.com/ec2/. Accessed Sep 23, 2013
[136] Walters MS (2005) NUBB: A Network Usage-based Billing System. In: Proceedings of the 33rd Annual Conference on User Services, ACM, New York, NY, USA, SIGUCCS '05, pp 426–429
[137] Wang D et al. (2011) Is Pay-per-click Efficient? An Empirical Analysis of Click Values. In: Proceedings of the 20th International Conference Companion on World Wide Web, ACM, New York, NY, USA, pp 141–142
[138] Wetterstrand K (2013) DNA Sequencing Costs: Data from the NHGRI Genome Sequencing Program (GSP). http://www.genome.gov/sequencingcosts/. Accessed Sep 23, 2013

Part II
Real-Time Data Analysis in Personalized Medicine

This part gives examples of how to design and implement specific applications enabling real-time analysis of scientific data. Furthermore, it provides guidelines to operate and to exchange huge data at fast pace. The following chapters are intended for researchers and medical experts, who require to work with big data on a daily basis. It also provides guidelines for IT experts how to operate on these data from a software engineering perspective.

Figure II.1 depicts the HIG system architecture as FMC block diagram focusing on IT components relevant for this part. For example, it shows specific extensions to enable cohort and pathway analysis, entity extraction to enable semantic search within unstructured text documents, and tools to combine and correlate data stored within the HIG system.

Table II.1 provides an easy navigation to identify relevant topics and target groups within this part of the book.

Title	Chapter	Audience
Real-time Analysis of Patient Cohorts	6	C, R
Ad-hoc Analysis of Genetic Pathways	7	C, R
Combined Search in Structured and Unstructured Medical Data	8	C, P, R
Real-time Collaboration in the Course of Personalized Medicine	9	B, C, P, R

Table II.1: Structure of Part II and addressed audiences (B = Bioinformaticians, C = Clinicians, P = Patients, R = Researchers).

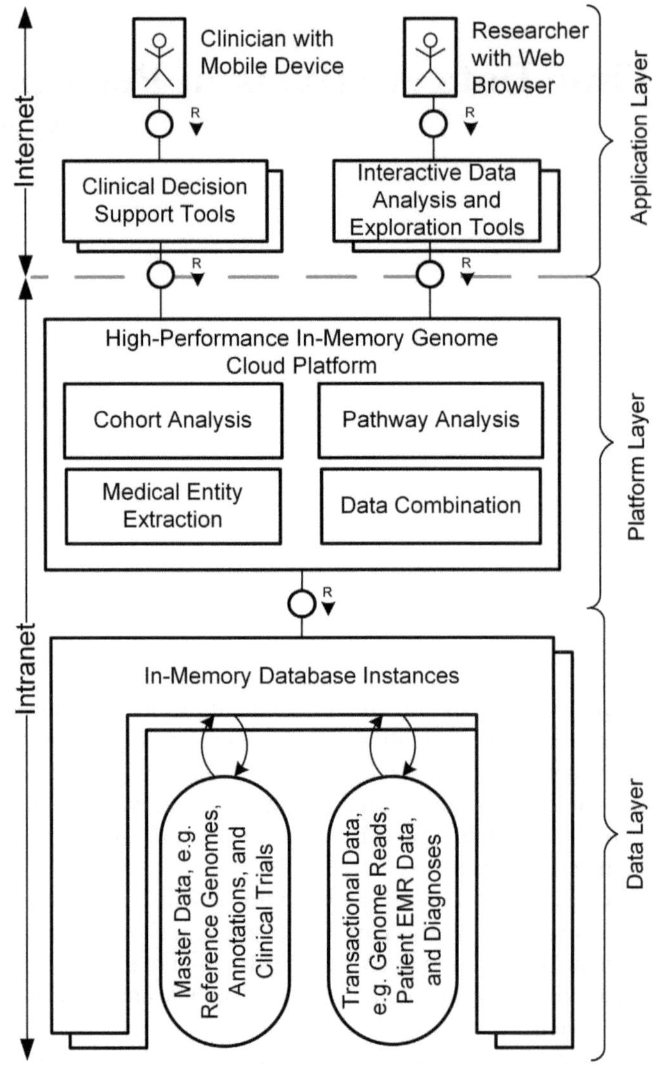

Fig. II.1: The HIG system architecture for real-time data analysis.

Chapter 6
Real-time Analysis of Patient Cohorts

Ricarda Schüler

Abstract In today's medicine, many complex diseases, such as cancer, are treated with a standardized therapy plan. However, this standard plan will usually not work in the same way for everybody. Researchers assume that this might be due to certain variants in patients' genes. Currently, researchers analyze large groups of patients with the statistical language R to investigate such cases further. But R is only a language and cannot store data, so that the data has to be loaded and often needs to be converted into the right format. In-memory databases could provide an alternative since they can store and handle a huge amount of data and allow analyzing them using the Structured Query Language. In this work, I propose to use the integration of statistical analysis in the database in a cohort analysis environment. I will show how an in-memory database can be used to analyze patient groups on the basis of k-means and hierarchical clustering. Therefore, I will compare the performance of clustering algorithms executed within an in-memory database and the statistics environment R. To give researchers the possibility to use the algorithms I developed a prototype, which provides a visualization of the clustering results and parallel clustering with several genes. In the future, it could be possible to perform a statistical analysis in the patient cohort analysis field using an in-memory database and save time for loading and preparing data. Thus, patient cohorts could be analyzed directly in the database without a system change.

6.1 Introduction

In course of bioinformatics, data analysis is important. In most cases there is one standard therapy plan suggested to use to treat one disease. But this standard therapy plan might work for the majority of the patients while failing for some remaining others [143]. Researchers found out that genomic variations could be a reason for differing treatment successes. But in many cases nobody knows what variation may have caused these differences.

To discover the relevant mutation, a patient cohort analysis can be performed. A cohort is defined as a group of people who share a characteristic over a certain period of time [144]. Hence, its analysis aims at showing certain characteristics concerning similarities and distinctions. This can be done by different statistical methods, e.g. clustering or regression.

Many scientists use the statistical language R to analyze data [159]. However, R loads its data from files and does not support real data manipulation. In contrast, databases are capable of aggregating data before an analysis, which allows researchers to obtain features of the structure of the data. Nevertheless, there is no possibility to perform advanced statistical analysis in the database. Therefore, the data needs to be loaded into another system such as R.

In this work, I will discuss how an in-memory database can be used to not only store and aggregate data but to execute cohort analysis. In order to do so, the method of clustering will be used to analyze patient cohorts. I will discuss the hypothesis that an in-memory database is as fast as the language R with data loading to do cohort analysis with clustering. Therefore, the usage of in-memory technology will be compared with the usage of R [150, Chap. R].

This work is structured as follows. In Section 6.2, I will describe the work being done in the field of cohort analysis and how hierarchical and k-means clustering work. The section also describes the utilized methods including the choice of in-memory technology, the connection between R and the in-memory technology, used data and implementation of the clustering algorithms. Section 6.3 shows what the developed system looks like and how it can be used. The benchmark setup and results are discussed in Section 6.4 and Section 6.5. The contribution concludes in Section 6.6 with a resume and outlook.

6.2 Related Work

In the following, the analysis of cohorts will be introduced. Additionally, two clustering algorithms will be described, which can be used in the process.

6.2.1 Cohort Analysis

Cohort Analysis can be performed using several techniques, such as regression or clustering. Another method was introduced by Zheng et al., who developed a tumor classification from gene expression data [163]. Hence, they just analyzed patient cohorts with so-called Independent Component Analysis (ICA), Eigen assay extraction and support vector machines. Furthermore, the usage of clustering algorithms is very common. For example, Sherlock as well as Jiang et al. describe how clustering algorithms can be applicable to a combination of gene data [158, 145].

Clustering algorithms can be used to discover relations between genes according to their functionality [140]. They can also be used to obtain a starting point for understanding the basic cell processes and what effects these and therefore what effect they have in humans, e.g. which genes are responsible for a regular cell death [140].

There are different kinds of clustering of cohort data. Jiang et al. have described these kinds and which clustering algorithms are useful in which case [145]. The following paragraph describes parts of the work of Jiang et al. and what has been used in my contribution. In the clustering context, objects are the variables, which should be classified into groups. The features are the properties of the used objects. Clustering can be distinguished into:

- **Gene-based clustering**: Genes are the clustered objects and samples are the features,
- **Sample-based clustering**: Samples are the objects and genes are the features, and
- **Subspace clustering**: Samples or genes can be both objects and features.

In this work, I will concentrate on sample based clustering. Nevertheless, the used technology is also applicable on gene based clustering. According to Jiang et al., sample based clustering is used to find relevant genes for a specific phenotype or to reduce the dimensions of characteristics to get the phenotype. A phenotype is the visual appearance of a person, which is characterized by environmental influences and the genotype, i.e. the genetic dispositions [155, Chap. 3]. This means sample based clustering helps, e.g. to find the reason for specific diseases.

The way of clustering the samples highly depends on the used data. If phenotype information is available, supervised analysis can be used, which means that possible relevant genes are known or can be extracted easily with the phenotype information, such as the particular disease a patient has. If there is no such phenotype information available unsupervised analysis must be used. This is more difficult because the relevant genes have to be extracted first without knowing the additional information [145]. As detailed in Section 6.4.1, the data used in my case includes such phenotype information. Therefore, in the following I will concentrate on supervised analysis.

According to Jiang et al., three steps are used for supervised clustering:

1. Selection of training samples,
2. Selection of informative genes, e.g. by neighborhood analysis approach, support vector machines or ranking based methods, and
3. Clustering of samples.

Chen describes a gene selection algorithm with ranking based methods in usage of gene expression data [141].

Because of the selection of the genes' conventional clustering algorithms like k-means clustering, hierarchical clustering, Self-Organized Maps (SOM) are applicable [158]. All of the algorithms previously mentioned need a decision how many clusters should be built [164, Chap. 16.1].

A clustering method described by Yeung et al. defines clusters and the number of clusters on the basis of mathematical models [161]. This method, known as model-based clustering, allows to finds a good but not necessarily optimal model and cluster number suitable to the data. The authors analyzed how this method can be used to cluster gene expression data. They used synthetic and real data. Being applied to artificial data, the algorithm finds the right model and cluster number. For real data, the method finds clusters similar to leading heuristic clustering algorithms and suggests a cluster number.

The aforementioned algorithms k-means clustering, hierarchical clustering and SOM are very common. The used in-memory technology provides these three clustering algorithms [157]. I decided to use two different clustering algorithms, the hierarchical one and the partition-based clustering method k-means, in order to cluster patient cohorts with genomic variants used as features.

6.2.2 K-Means Clustering

There are a variety of clustering algorithms available [160]. All of them have special purposes and disadvantages. The k-means clustering algorithm is very common and used in many fields, like image processing, sales forecasting or analysis of gene expression data [149]. The calculation steps according to MacQueen are:

1. Determine random cluster centers or choose predefined centers,
2. Sort values to nearest cluster center, with the nearest cluster center defined by a distance function, e.g. Euclidean or Manhattan,
3. Recalculate cluster centers as barycenter of previous clusters, and
4. Repeat steps two and three until the maximum iterations are reached or the change of the cluster centers is under a given threshold [151].

This cluster algorithm needs the number of clusters as input parameter, so that the user will know the most suitable number for the investigated data. As Zvelebil and Baum explain, this might be difficult in most cases of genome data analysis [164, Chap. 16.3]. They suggest a calculation with different numbers of clusters. Often the predefining of initial cluster centers is not possible but choosing random initial cluster centers can provide different results for every calculation. The input dimension should not be too high because the algorithm needs a long time to calculate the nearest cluster center of each value [145].

As described above, each point will be sorted in a cluster, followed by the calculation of the new cluster centroids. This leads to a susceptibility of noise of the input data because non-relevant values influence the clusters in the same way as the relevant ones [145].

According to Jiang et al., the time complexity of the k-means algorithm is $O(l * k * n)$, where l is the number of iterations, k the number of clusters and n the number of input values [145]. Zhao et al. demonstrate that k-means clustering can be sped up linearly to the number of computers up to a certain point with the usage of map reduce [162]. The k-means algorithm can also be calculated in parallel in some parts. The distance calculation between the values and cluster centers can be calculated for each value in parallel. The next step, computing the new cluster centers, can be performed for each center in parallel. I expect that if all centers and distances can be calculated in parallel, the time complexity approaches $O(l)$.

6.2.3 Hierarchical Clustering

Hierarchical clustering is also a widely used clustering algorithm, which is used to see the hierarchical structure of a data set. The input is a $N \times N$ distance matrix, where N is the number of values to cluster [146]. In this matrix, the distance between the values is defined. To calculate such a distance, matrix functions such as the Euclidean distance, the Pearson correlation coefficient measure, or the Manhattan distance are available [164, Chap. 16.3]. The following paragraphs describe some distance functions for gene expression data according to Zvelebil and Baum [164, Chap. 16.2].

The Euclidean distance function calculates the distance between two points that are represented as vectors in n-dimensional space. It considers the direction and the size of the distance. The Euclidean distance is a method in which the priority of the N different values is not regarded. For example, a weight difference of 1 kg has the same priority as a temperature difference of 1 °C. This effect will likely result in a non-realistic analysis of the data and therefore produce meaningless information. If the data parts have the same priority, the algorithm can calculate the distance quickly.

The Pearson correlation does not measure the distance based on the real values, but by using the shape of the pattern in a dataset. It calculates the similarity of the direction of two expression points after normalizing the vectors [158]. In most cases, this gives better results than the Euclidean distance. However, it does not give respect to the fact that, e.g. highly expressed points are more important than poorly expressed ones [164, Chap. 16.2].

After defining the distance between two objects, the clustering steps according to Johnson are:

1. Define each value as a single cluster,
2. Select the two nearest values to build one new common cluster,
3. Merge the next nearest cluster, which can be calculated with different methods, e.g. average centroid method, and
4. Repeat step three until the desired amount of clusters is reached [146].

The result of the algorithm is a tree representation, a so-called dendrogram, in which the proximity and therefore relations between the objects are shown [164, Chap. 16.3]. This is a relevant difference to the previous described k-means clustering, where only clusters were defined without showing exact relations between the clusters.

A disadvantage is that an unfavorable decision at the early steps of the algorithm cannot be corrected thereafter [145]. Hence, a merging of two clusters that do not belong close together is irreversible. The previously described k-means clustering will prove its previous decisions on every calculation step.

The time complexity of hierarchical clustering is $O(n^2 \log n)$ where n is the number of input values [145]. This means that k-means clustering scales better with big data than hierarchical clustering. Therefore, k-means clustering is preferable if the speed is most important.

6.2.4 In-memory Technology Building Blocks

In a calculation scenario with the programming language R, the system has to load the data from the database via SQL or directly from hard disk, e.g. from a Character Separated Values (CSV) file. In case a disk-based database is used, the database must also read the data from a hard disk.

With in-memory technology, the data is loaded into memory once and resides there for use by algorithms, aggregations, and queries. Analytical algorithms like clustering can be implemented inside the database where they have direct access to the data.

Today's availability of servers with very large main memory, makes it possible to handle big data in-memory [154]. Also, the speed of data access increases when using main memory instead of the hard disk. As described in Section 1.4.3, lightweight compression techniques can further reduce the memory footprint of collected data while allowing calculation and queries on the compressed data.

A trial containing all German patients diagnosed with cancer within a year would include information about 480,000 people [147]. One patient could have up to 3 GB of genomic data. Assuming that only the protein-encoding genes are sequenced, around two percent of the 3 GB of data is used. To analyze this amount of data, the reading time from disk alone takes about ten days. At this point, analysis would still need to be executed. This period of time avoids a fluent interaction or even a real-time analysis. To read this amount of data from main memory, only about two hours are needed. Therefore, a utilization of in-memory technology will speed up the analysis of genome data, if the algorithm can be used directly on the data loaded into memory. In theory, this saves more than nine days, operating with the same algorithm. This represents a speedup that allows the user to interact with the calculation platform directly.

The next paragraph describes the parallelization of in-memory technology [154]. In an operating system, threads need to be mapped to physically available hardware resources. A Central Processing Unit (CPU) can be considered as such a single hardware resource. Therefore, optimal parallelization can be spoken of when each CPU can execute one part of an algorithm or query. Another point worth mentioning is the used resources. If the single queries or algorithm parts need the same data, the parallelization is limited. Thus, partitioning of the data into different parts can be helpful.

In cohort analysis, the amount of available information for a patient usually varies. Additionally, not all of the information is used for each of the clustering iterations. The column store of the in-memory database enables the reading of only those columns that are required for the algorithm or query, as explained in Section 1.4.1 [154].

6.3 Application Example

This chapter describes the setup of the developed system and the user interfaces.

6.3.1 Architecture

The research results are integrated in the High-performance In-memory Genome (HIG) platform, which is described in Section 1.5. The three layers – data layer, platform layer, and application layer – are adhered to, which is shown in Figure 6.1.

The data layer contains the patient cohorts data, as detailed in Section 6.4.1. The platform contains stored procedures, which calculate the clusters with the help of the SAP HANA Predictive Analysis Library (PAL) or with the help of R. Overall, there are four different clustering methods: k-means clustering with R as well as with in-memory-technology and hierarchical clustering with R and with in-memory technology. In Section 6.4.2, the usage of the existing algorithms is described. The connection between R and the in-memory database is detailed in Section 6.4.3.

The application layer is responsible to integrate the HIG frontends for web and tablet devices. These two applications present the results of the clustering. More applications can be easily added to show the results. The following part will show what the provided applications look like and how they can be used.

6.3.2 Application

The integration into the existing web UI of the HIG platform is depicted in Figure 6.2. The targeted user group of the developed prototype is researchers. After selecting this tab, the user can choose the trial that should be analyzed. The number of clusters and the clustering algorithm – k-means or hierarchical clustering – can also be selected. Afterwards, the scientists can either:

- Select three genes and positions, which might have a significant influence during clustering, or

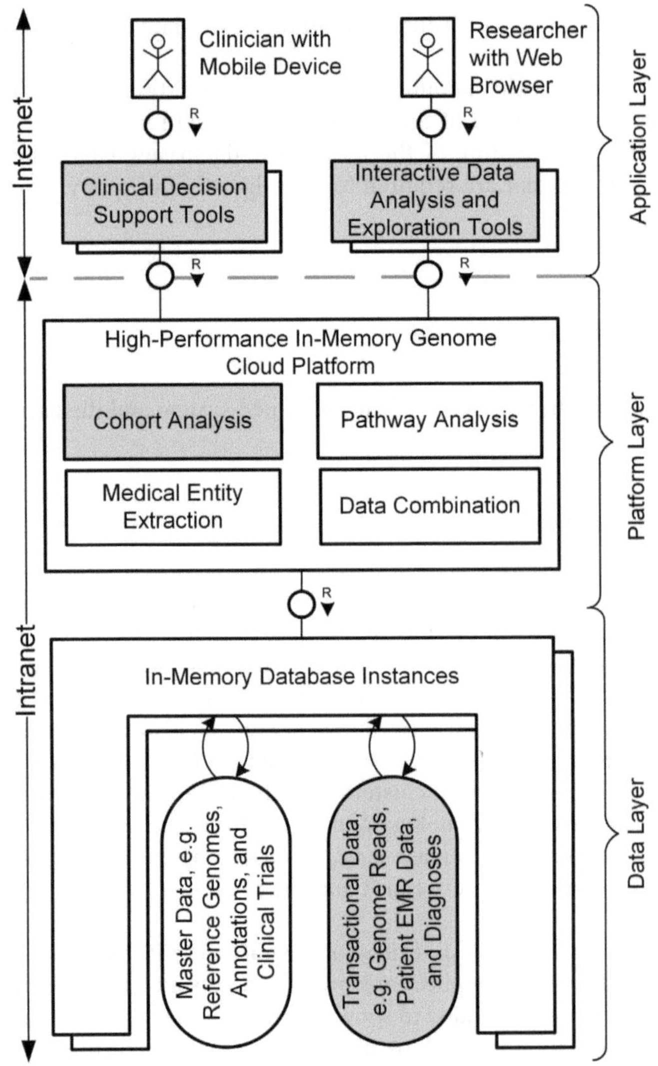

Fig. 6.1: Architecture of the prototype based on HIG architecture modeled as FMC block diagram.

- Choose to calculate in parallel 20 combinations of the most mutated genes in the data set.

As a hint for the user, when working with the first option, the ratio of how many people of the selected cohort have a mutation on this gene is shown for every gene. Additionally, the proportion of people having a mutation at the exact position compared to the people who have a mutation on the whole selected gene is given. This additional information not only helps the user to select suitable genes and

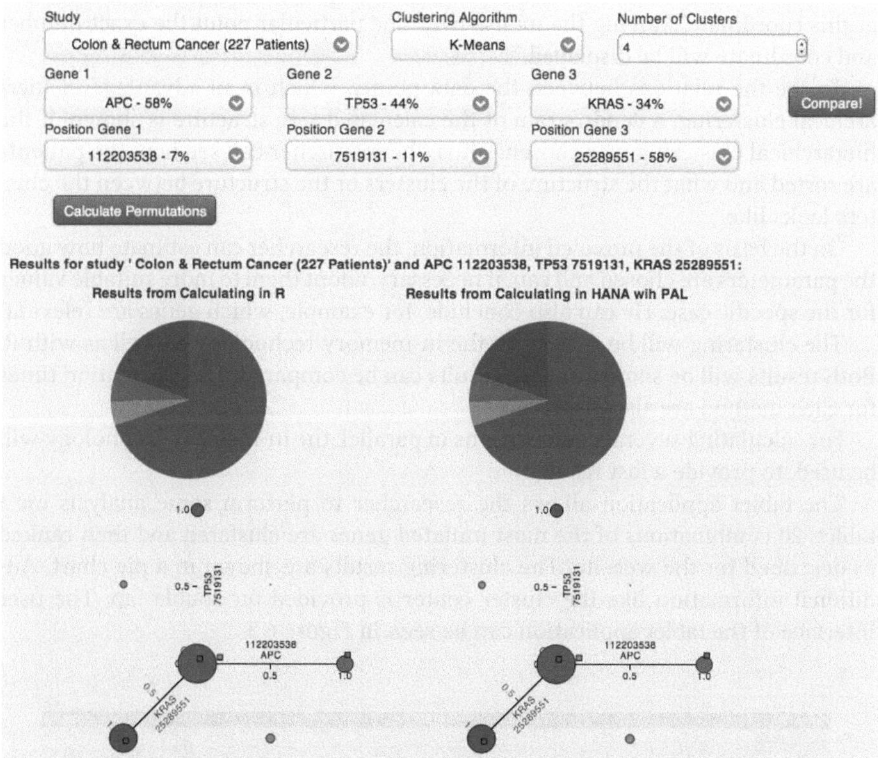

Fig. 6.2: Website interface of the prototype, which is used to cluster patients based on their genotype data with in-memory technology.

positions, but also gives the scientist the opportunity for educated guessing based on his very own knowledge and experience.

The second option provides a ranking of the used gene-position pairs starting with the biggest cluster that does not include patients who have no mutation at the selected genes.

The cluster results are presented in different ways depending on the user needs. In the pie chart, the clusters with their number of patients are shown. This helps the scientist to understand, which mutation combination is quite common in the selected cohort. When moving the mouse over a particular item, additional information like the coordinates of the cluster center and the number of people in the cluster will be shown.

With the information presented in the pie chart the user is not able to know how well the found cluster fits to the used data. Therefore, a three-dimensional scatter plot is given, which includes cluster centers and the patients with their mutations as features. If many people have the same coordinate, only one data point will be drawn, while the size of the point scales with the amount of people at this coordinate. Moving the mouse over one particular point the exact number and coordinate will be displayed.

To see the relations between the data points, which is an advantage of hierarchical clustering, a dendrogram of the calculated tree structure is shown, if the hierarchical clustering was chosen. Thus, the researcher can see how the patients are sorted and what the structure of the clusters or the structure between the clusters looks like.

On the basis of the provided information, the researcher can estimate how good the parameters are chosen and can, if necessary, adopt them to more suitable values for the specific case. He can also conclude, for example, which genes are relevant.

The clustering will be done with the in-memory technology as well as with R. Both results will be shown, so that results can be compared. The calculation times for each method are also displayed.

For calculating several combinations in parallel, the in-memory technology will be used, to provide a fast result.

The tablet application allows the researcher to perform some analysis on a tablet. 20 combinations of the most mutated genes are clustered and then ranked as described for the website. The clustering results are shown in a pie chart. Additional information like the cluster center is provided on double tap. The user interface of the tablet application can be seen in Figure 6.3.

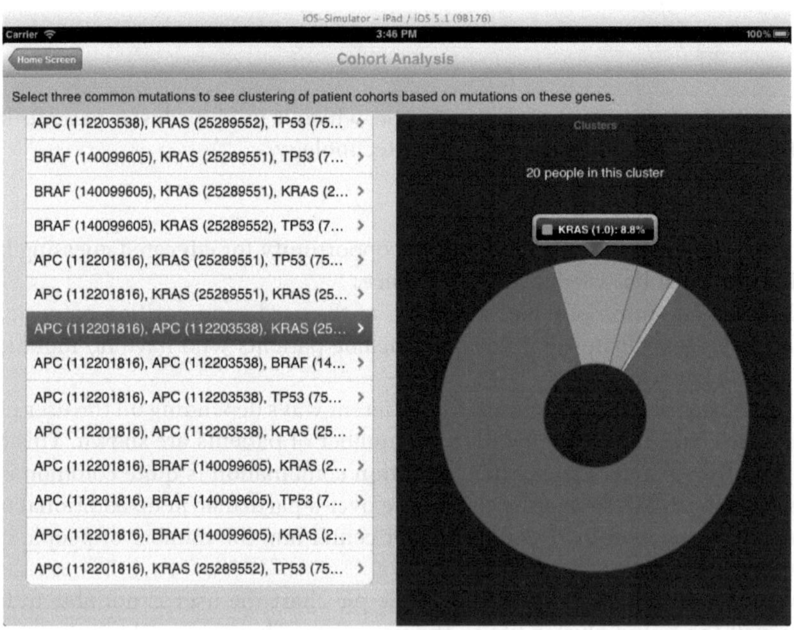

Fig. 6.3: Mobile user interface of the cohort analysis, which allows clustering several genes in parallel and visualizes the results as charts.

6.4 Benchmarks

I will show that an in-memory database is at least as good as R with regards to the speed of cluster calculation with k-means and hierarchical clustering. The utilized algorithms are essentially the same, but I expect that the in-memory database can speed up the calculation through parallel execution of certain parts. Additionally, in the IMDB loading the data is not necessary while R has to load them additionally for the calculation.

In this chapter, I describe the setup of my conducted benchmarks.

6.4.1 Benchmark Data

I tested the two clustering algorithms, k-means and hierarchical clustering, on the basis of real trial data. The data comes from several trials, e.g. the trial "Comprehensive molecular characterization of human colon and rectal cancer", which contains 276 patients with additional information, like their age at the time of diagnosis, kind of disease and number of relatives with the same disease [152]. The study also includes partial genomic data. This means that only the patient's mutations relevant for the purpose of the trial are shared. The data consists of a total of 90,059 mutations and has a size of 40 MB in the CSV format as opposed to 15 MB in the database. This represents a compression rate of 8:3.

Table 6.1 provides an overview of the studies used. As can be seen, I used three additional studies with different cancer types. These studies include the same information as described above.

Trial Name	#Patients	#Mutations	Size [MB]
Comprehensive genomic characterization of squamous cell lung cancers [153]	178	65,305	62.5
Integrated genomic characterization of endometrial carcinoma [148]	248	182,824	75.7
Comprehensive molecular characterization of human colon and rectal cancer [152]	276	90,059	40.0
Integrated genomic analysis of ovarian carcinoma [139]	316	19,356	10.8

Table 6.1: Study data used for evaluation.

The original data has string values as patient identifiers, which are used as join attributes or as identifiers in the clustering algorithm. My own experiences yield that a join on string attributes is noticeably slower than a join on integer values. Thus, I replaced the alphanumerical identifiers with numerical ones during the data import, in order to speed up my queries.

Because of the small number of patients, the amount of data is not sufficient to test the performance of the clustering algorithms. On the basis of the described data

I generated new data. To ensure realistic data, the same distribution of mutations and the same number of mutations per participant has to be kept. Therefore, the easiest solution would be a duplication of patients. In reality, such data distribution would not occur. It can also increase the amount of caching during the calculations and distort the measurements. Thus, I generated the new data with nearly the same distribution of the three most mutated genes. The other mutations of a patient are selected randomly from an arbitrary pool of possible mutations. The number of mutations of a patient is determined by the average and standard deviation of the number of mutations of the participants of the underlying study. This procedure results in a data set of patients containing the same distribution of real data in some selected points while also having some natural noise.

The data will be stored in an in-memory database with three tables per study. One table contains data such as age and diagnosis of the participants, another one the mutations of tumor samples, and a third one holds the mapping between the samples and the people.

The k-means and hierarchical clustering algorithms of my test system require their input as database table with the following columns: ID, attribute 1, attribute 2, ..., attribute n [157]. Therefore, the data has to be converted into this format. In my case, possible attributes are gene-position pairs, e.g. gene KRAS and position 25289551. The value of every single one of these attributes can be either "0" or "1" for each patient, where "0" symbolizes that the patient has no mutation at this position and "1" stands for a mutation. Every patient is represented by one table row. In the example depicted in Table 6.2, three attributes are used and e.g. the patient with ID 15 has mutations on the KRAS and TTN genes, but no mutation on APC.

ID	KRAS, 25289551	APC, 11220353	TTN, 179320989
15	1	0	1
18	1	1	0
21	0	0	1

Table 6.2: Example of an input table for clustering algorithms.

Every gene-position pair is one dimension in the clustering algorithm. This means that a data set with full genome samples has three million dimensions. But the clustering algorithms do not work with such a high number of dimensions. The solution is a preselection of the genes.

Selecting the relevant genes is a complex problem. It is only possible if some additional data to the gene expression data is available (supervised selection), such as the kind of disease, as mentioned in Section 6.2.1 [145]. In this work, the gene selection will not be investigated. For simplification, I use the genes with the largest numbers of mutations in the data set.

6.4.2 Benchmarks for In-memory Technology

The following experiments were conducted on a prerelease of SAP HANA database, version 1.50.00, built on July 26, 2013, as a concrete implementation of the in-memory database technology [142]. It contains the latest development version of PAL, which provides the k-means and hierarchical clustering [157]. This section will describe how I used these algorithms to cluster the data described in Section 6.4.1.

The output of the k-means clustering is a mapping from each patient to a cluster along with additional information for each calculated cluster, like the center coordinates. This information can be used to understand how the data is structured.

The two algorithms allow the adjusting of different parameters, e.g. the cluster number, thread number or maximum iteration size [157]. Among other parameters, this is used to execute benchmarks with different settings as described in Section 6.4.4. Table 6.3 depicts additional parameter settings for k-means clustering that were set with fixed values and not modified during the benchmarks.

Parameter	Value
Init type	2
Distance level	2
Exit threshold	1×10^{-6}
Normalization	0
Maximum iterations	1,000

Table 6.3: Parameters used for the k-means clustering in PAL.

The init type specifies the definition of the initial cluster centers. In my case, 2 - random number with replacement is used. For the distance level, which defines how the distance between an item and a cluster will be calculated, the Euclidean distance was chosen. Furthermore, I do not use any normalization. The algorithm performs 1,000 iterations until the final cluster result is reached. Only if the threshold defined above is reached (almost no changes in calculation of the new cluster centers), the algorithm will stop earlier. The output consists of all merging steps as well as the mapping of the patients to one cluster.

I chose the Euclidean distance function, because the disadvantage of non-priorized input data described in Section 6.2.3 is not relevant for this particular application. The used cluster method is the so-called single linkage clustering. It defines the proximity of two clusters with the minimum distance between two members in the clusters [164].

K-means clustering as well as hierarchical clustering can be performed in the in-memory database with different input data in parallel. This can be necessary for medical researchers, if they do not know exactly, which genes or gene combinations are relevant. It gives the possibility to test some combinations and find the most relevant ones.

6.4.3 Benchmarks for R

As a comparison to the algorithms of the IMDB, the widely used statistical language R is used, which can be connected with an IMDB in multiple ways:

- Inside R: JDBC connection or using RHANA package, and
- R server architecture: with TCP/IP or shared memory.

The inside R solutions are called from an R console. The JDBC connection uses Java as basis. A positive fact is that existing R code can be reused and new R code is written in the same way that R users are used to. The new code can be used with other data in the same way. When creating new tables in the IMDB, this solution does not support all of the in-memory table structures. The RHANA code supports

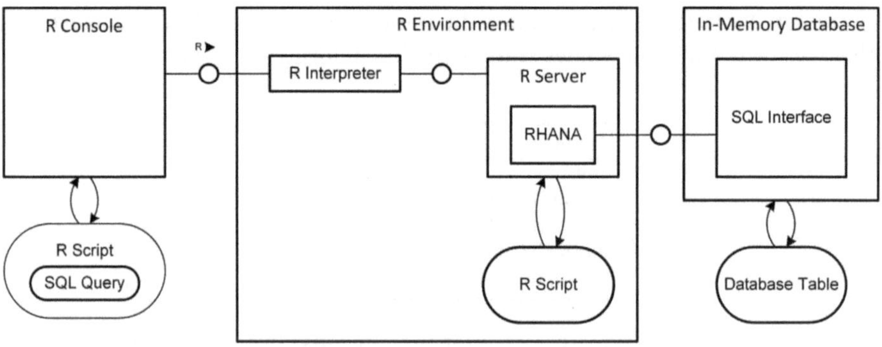

Fig. 6.4: Architecture of RHANA interconnecting R and the in-memory database according to [157] modeled as FMC block diagram.

the data handling that R users are familiar with as well as all IMDB database table structures. However, not all of the R objects can be transformed easily into database tables. The architecture, which is depicted in Figure 6.4, includes the usage of the R interpreter. It also uses an R server, which allows a fast data transfer directly from the database to the R interpreter. The data has to be transformed into the R data structure `data.frame`, which can then be used, e.g. to cluster the data.

The R server architecture allows the user to handle R code like stored procedures and be called from other database connections easily. The transfer between the database and R is not supported for all R data types. This means that data preparation needs to be done in R instead of doing it in the database. This handling will cost more time. The data has to be transferred to R and after calculation the results have to be transferred back to the database. Hence, for a TCP/IP connection based solution the data stored in a columnar form, needs to be transformed back to a row-based data structure. As shown in Figure 6.5, the IMDB starts the data transfer via TCP/IP. The R system can be located on another machine.

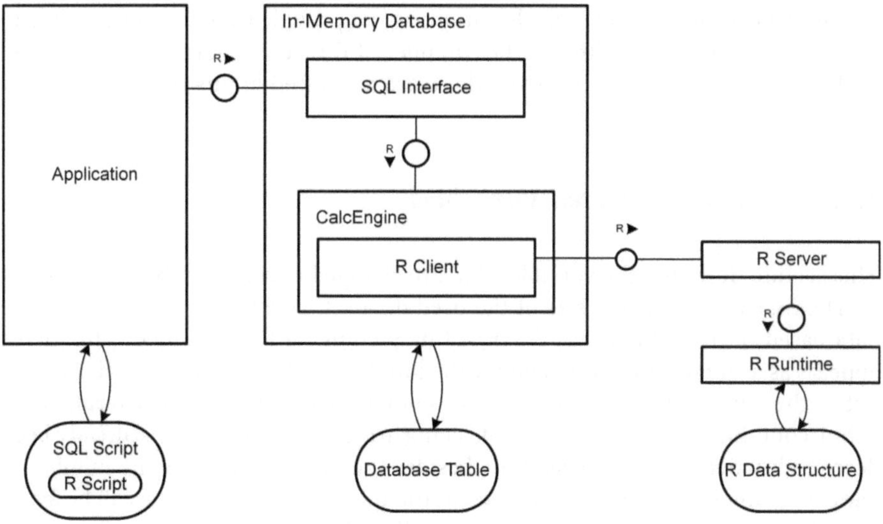

Fig. 6.5: Architecture of R client embedded in IMDB modeled as FMC block diagram based on [157].

Even in a shared memory solution, the data has to be transformed into the right format. The database and the R system have to be located on the same machine, otherwise a shared memory cannot be used. A speed up by a factor of two is estimated in contrast to the TCP/IP connection for the data transport.

I decided to use the R server architecture because it is necessary to reach the R code from a web service, which will be described in Section 6.3.1. The database has to be reached anyway, so that a call of R via the database is preferable. To compare the calculation times of R and PAL, the R instance and the in-memory database should run on the same machine. As the used system was a distributed system, a solution with shared memory was not possible. Instead, the R server architecture with data transferring via TCP/IP was chosen.

The input for R has to be structured in a certain way, as depicted in Table 6.2. But it has to be converted into the R data type data frame first. I used the provided method `kmeans`, which is included in the package `cluster`. As internal algorithm the Lloyd implementation is chosen, as it is the same algorithm that was implemented in the IMDB [157].

The hierarchical agglomeration is calculated with the `hclust` function. This algorithm needs a distance matrix as input, which is computed with the function `dist` and the Euclidean distance method. The `method` parameter of the `hclust` is set to single linkage. The output of this algorithm is a tree of the input objects, which can be displayed in R very easily with the `plot` function. Besides the tree, there is information like the merging steps. This is very similar to the output of the in-memory calculation. The tree is cut at a specific cluster amount with the function `cutree`.

Parallel execution inside the algorithms is not provided, so it is not possible to speed up the algorithm with a higher number of threads. In some cases, there are packages to calculate in parallel, but these are not tested in this contribution.

6.4.4 *Impact of Selected Variables*

This section describes the variables that are changed during each benchmark.

The first variable is the size of the used data set. Due to the generation of the data based on real studies, as described in Section 6.4.1, the number of patients appears as a suitable measurement for the input size. The generated database view out of the study data, which is used as the input matrix for each clustering algorithm, contains the number of patients being arranged in a row as has already been described in Section 6.4.1. Therefore, besides the number of genes the number of patients defines the required data size. In the future, even more data will be available. Assuming that it will be possible to collect data of all incidences of cancer in Germany within two years, about 1,000,000 patients will be available [147].

The second variable is the number of genes used as input. Both systems (R and PAL) will be slower with more dimensions. It would be a great advantage for researchers if one system can scale in a better way. If it is possible to use more dimensions, the clustering has a greater scope, so that more relations and a larger context can be detected.

Only a single variable is changed for each benchmark while the remaining parameters are set according to Table 6.4.

Variable	Value
Number of genes	3
Number of patients	744

Table 6.4: Standard values for each benchmark measurement.

6.4.5 *Test Procedure and Technical Environment*

This section describes the procedure of the benchmark tests, the steps being:

1. Create a new database view for the data,
2. Set parameters,
3. Execute k-means clustering with R and PAL,
4. Execute hierarchical clustering with R and PAL, and
5. Calculate average and standard deviation.

Prior to each time measurement, an input table will be created, in which the data mapping will be done as described in Section 6.4.1. It is necessary in most cases because the number of genes defines the database view. A consequence is also that the used data is already loaded into the main memory, so that the full speedup of the IMDB can be used.

Items 3 and 4 are executed several times until the standard deviation is less than five percent. The first measurement is not used for benchmarking because this time is significantly higher than the following measurements for a small data size. The loading of the data causes this.

The time is measured with Python using the `time` module and the `time` method. Hence, for every calculation a network time is included but every measurement has the same delay. This means that the result can be compared without taking into account different time delays. The time consumed by the clustering in R is measured directly in the R code, which is executed on the R system. I used the method `proc.time`. This time gives an impression how fast the real calculation in R is.

The last step is the calculation of the average and the standard deviation of all measurements with different parameters.

I executed the benchmarks on a distributed system including two hosts. Each host has the hardware listed in Table 6.5.

Processor	8 x Intel® Xeon® E7-8870 @ 2.40 GHz
Main Memory	64 x 32 GB DDR3 1066 MHz + 64 x 16 GB DDR3 1066 MHz
Operating System	SUSE Linux Enterprise Server 11.2 (Kernel 3.0.13-0.27-default)

Table 6.5: Configuration of benchmark system.

The in-memory database used these two hosts, while R was running on only one of the hosts because running on a distributed system was not possible. The R application, which is used for comparison, has the version 2.15 and was compiled by me. The Python version used for executing the benchmarks was 2.64.

6.5 Results and Discussion

In this chapter, I will present the results of the benchmarks, which were executed as described in Section 6.4. I will also evaluate and discuss the series of benchmarks and the conclusions that can be drawn according to the hypotheses.

6.5.1 Data Size

The clustering algorithms were tested with several data sizes. This gives an impression of the scalability with the amount of data of PAL in comparison to R. This scalability depends on the used data types and data handling in the algorithm.

To test this, the data was generated as described in Section 6.4.1 and was used with the setup described in Section 6.4. For each data set, a new study with a certain number of patients was generated. Hence, all of the data sets have the same structure and distribution, so that comparability is given.

#Patients	Mean I [s]	SD I [%]	Mean II [s]	SD II [%]
33	1.285	1.282	0.715	6.158
78	1.354	2.049	0.786	1.753
372	1.356	1.366	0.764	1.090
744	1.417	1.401	0.833	3.066
3712	2.365	3.366	1.751	4.660
7406	3.927	4.213	3.073	2.339
37187	15.166	2.321	13.460	2.092
74745	29.333	2.415	25.969	2.024
372498	152.919	3.443	136.966	2.041
744704	328.005	3.340	278.675	3.384

Table 6.6: Results of benchmarking with different data sets for k-means clustering with PAL and R (Experiment I = R, Experiment II = PAL, SD = Standard deviation).

Each series of measurements was performed until the standard deviation reached a six percent threshold. The internal times in R have a bigger standard deviation. This is a result of the measuring inaccuracy of the time function used there. The averages and standard deviations of k-means clustering with PAL and R are shown in Table 6.6. As visible in the table, the PAL algorithm needs less than one second for up to almost 1,000 patients, which allows an interactive exploration of the data. However, R always needs more than one second.

In Figure 6.6, the calculation time of the k-means clustering depending on the patient number is shown. The more patients are in a study, the more time the calculation needs. The points are again connected with a straight line to guide the eye. R and PAL are scaling in a linear way. This is the expected result, because this time the algorithm has a time complexity of $O(n * l * k)$ where the number of clusters (k) and the number of iterations (l) were constant, compared to the number of patients (n).

PAL is always faster than R, which is likely the result of the efficient data handling. PAL is in average 42.8 % faster than R concerning the values less than 1,000 patients and about 16.0 % faster with more patients. Especially the difference in calculating clusters saves the user a lot of time in total. Up to a number of 50,000 patients, the k-means clustering in PAL needs less than 30 seconds. This is an impressive result, because it shows that the in-memory database is a beneficial alter-

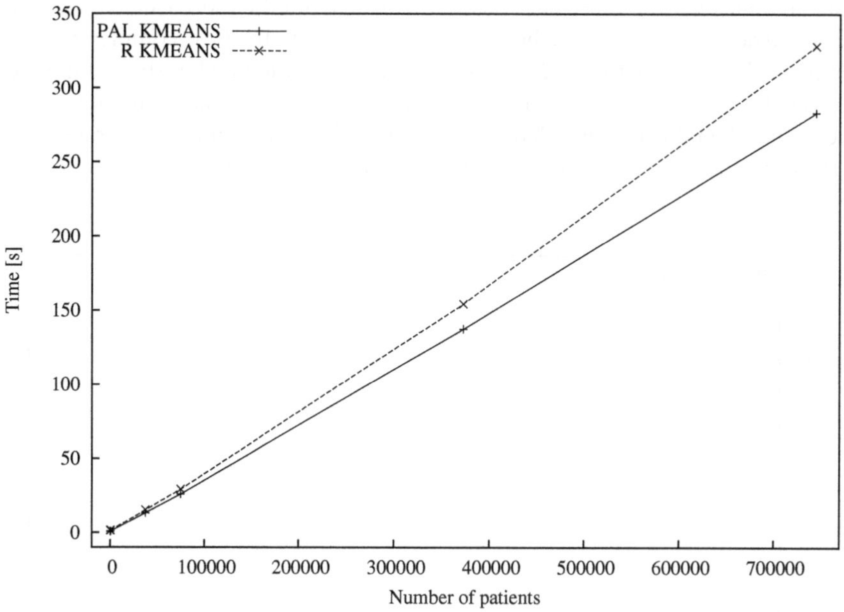

Fig. 6.6: Performance of k-means clustering with R and PAL dependent on the number of patients in the study.

native to R if much data has to be handled and when after the calculation some enrichment and further calculation with other data is desired.

#Patients	Mean I [s]	SD I [%]	Mean II [s]	SD II [%]
33	1.406	3.397	0.912	8.394
78	1.459	3.330	1.009	4.918
372	1.470	3.473	1.247	3.780
744	1.601	2.062	1.735	2.653
3,712	3.879	2.922	6.085	3.949
7,406	10.902	3.039	13.208	2.615
18,666	53.110	3.376	41.709	1.907
37,187	216.158	7.781	116.406	4.248

Table 6.7: Results of benchmarking with different data sets for hierarchical clustering with R and PAL (Experiment I = R, Experiment II = PAL, SD = Standard deviation).

Table 6.7 shows the results of the hierarchical clustering of PAL and R. It is noticeable that PAL is faster than R, especially considering the calculation of the maximum data size. Figure 6.7 demonstrates that the bigger the data size is, the longer the calculation of the clusters needs. This is not surprising because the more patients are analyzed the more data has to be loaded and the more calculation steps have to be done. I expected the behavior to be similar to the time complexity of hierarchical clustering, which is $O(n^2 \log n)$. The times of R matches this complexity, whereas the curve of PAL increases more slowly. The way a large amount of data is being handled during the calculation, e.g. based on the compression rate in the database, is one reason for this increase in calculation speed.

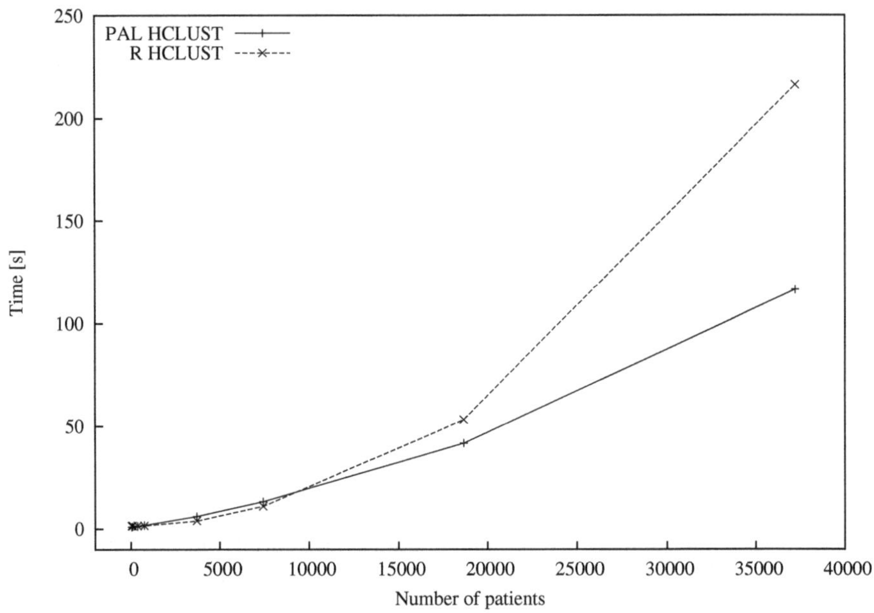

Fig. 6.7: Performance of hierarchical clustering with R and the in-memory database dependent on the number of patients in the study.

In Figure 6.8, the performance of the PAL algorithms for k-means and hierarchical clustering is compared. It shows that k-means is faster than hierarchical clustering. Hierarchical clustering calculates the distance matrix between the clusters existing in each iteration. The number of clusters reduces each iteration by one. Calculating the distance matrix between clusters is more expensive than calculating only the distance between two points, which k-means does for every calculation. Apart from the absolute time, the time complexity of the algorithms show that the execution time of k-means clustering increases more slowly than the calculation time of hierarchical clustering.

The hierarchical clustering aborted the operation when trying to cluster more than 40.000 people whereas k-means can handle more data. This is the result of the algorithm's structure because a distance matrix between all patients is calculated in the beginning. This matrix has a dimension of 40.000 x 40.000 and therefore has a space complexity of $O(n^2)$. K-means only has a space complexity of $O(n)$. The main

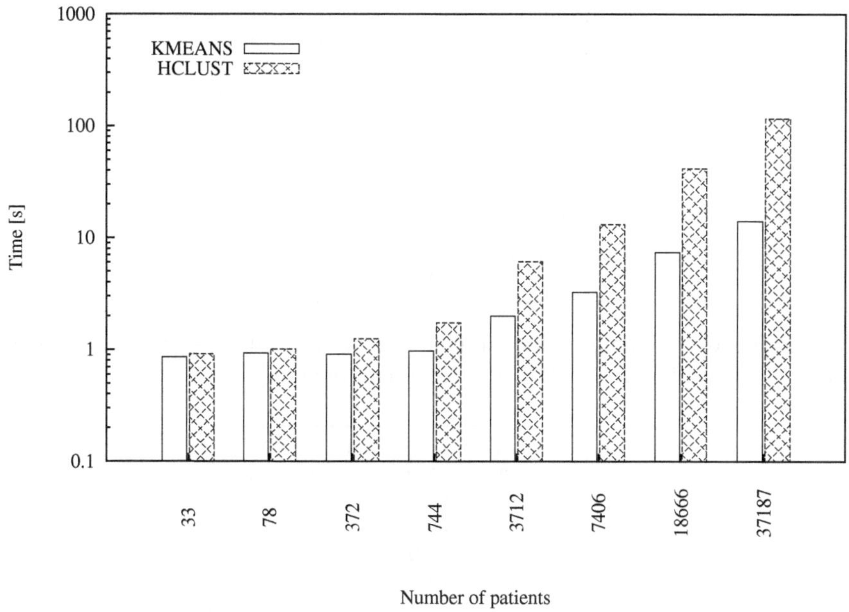

Fig. 6.8: Performance of hierarchical clustering and k-means clustering dependent on data size in comparison.

problem in PAL was the wrong usage of data types. This bug has been reported to the developers and will be corrected. This will hopefully shift the capability of the algorithm to a similar level than k-means clustering has.

6.5.2 Number of Genes

This section will show how the number of genes, and therefore the number of dimensions, has an impact on the calculation time of hierarchical clustering and k-means clustering.

The higher the number of genes that can be used, the more relations between genes can be inspected. On the other hand, it gets more difficult to detect these relations between the additional noise. In Table 6.8 and Table 6.9 respectively, the average results and standard deviations of the series are given. The standard deviation is always below seven percent, so that the measurements have a high quality.

In Figure 6.9, the calculation depending on the number of input genes is shown. The lines between the values are drawn to guide the eye. It is obvious that the in-memory calculations are much faster than R even for a high number of input genes, whereas the PAL curves seem to be constant. The PAL times will likely grow with more genes because it influences the algorithms for calculating the distances

#Genes	Mean I [s]	SD I [%]	Mean II [s]	SD II [%]
3	1.571	3.697	0.972	6.370
4	1.715	2.504	0.964	5.362
8	1.907	5.073	0.990	3.999
10	1.998	1.668	0.988	5.106
20	2.356	4.170	1.011	3.272
50	3.751	4.735	0.965	5.505
100	7.278	6.295	1.002	3.998
150	13.430	4.338	0.994	5.469
200	19.803	3.581	1.003	3.732
246	27.045	3.342	0.999	4.333

Table 6.8: Results of changing the number of genes for k-means clustering (Experiment I = R, Experiment II = PAL, SD = Standard deviation).

#Genes	Mean I [s]	SD I [%]	Mean II [s]	SD II [%]
3	1.601	2.062	1.735	2.653
4	1.740	1.959	1.747	2.068
8	1.965	4.324	1.729	2.785
10	2.055	3.971	1.739	3.336
20	2.416	4.472	1.769	1.662
50	3.789	2.016	1.764	1.756
100	7.176	4.135	1.756	3.218
150	12.846	4.802	1.758	2.953
200	19.110	3.940	1.775	4.218
246	26.466	2.878	1.792	3.132

Table 6.9: Results of changing data sets for hierarchical clustering (Experiment I = R, Experiment II = PAL, SD = Standard deviation).

between the patients and the cluster center or, in case of hierarchical clustering, between the clusters. It was not possible to test the algorithms with more dimensions, because the input data table was realized with a data view of the original data. This view contains one column for each dimension, as described in Section 6.4.1. To create this view, a transposition of parts of the input table was realized using joins. SQL in the in-memory database has a limitation of a maximum parse tree depth of 255 [156], which was reached in this case.

The in-memory database is a good choice to cluster when using a high number of dimensions because the benefit in performance is significant. In summary, PAL is a good alternative for clustering patient cohorts and provides similar statistical functions as the corresponding R implementation.

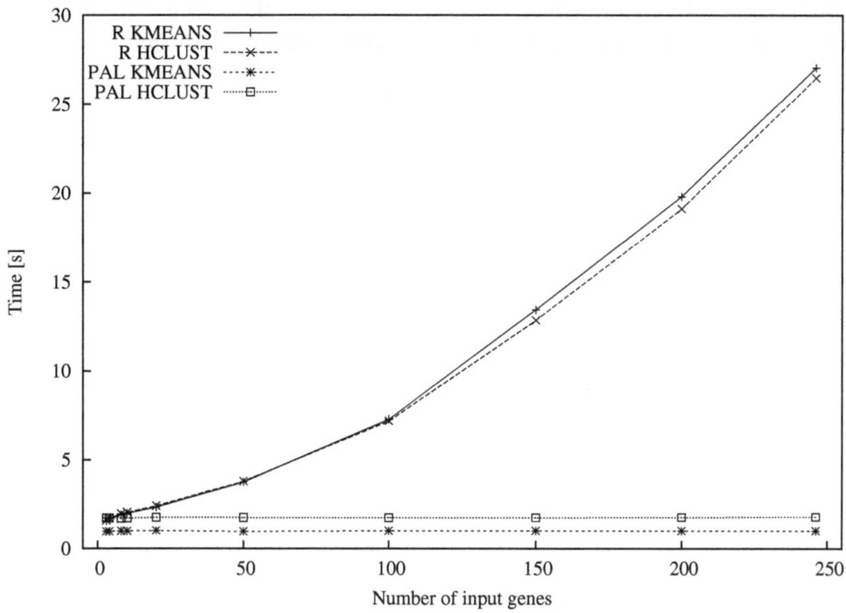

Fig. 6.9: Impact of number of input genes.

6.6 Conclusion and Outlook

I analyzed how well an in-memory database can be used to perform cohort analysis with hierarchical and k-means clustering. I compared the clustering algorithms provided by the PAL of the in-memory database with the clustering algorithms of the statistical language R. The basis for comparison was data interpolated from real data by duplicating under retention of the statistical properties as described in Section 6.4.1. During my benchmarks, I considered the number of input genes and the number of patients in a study.

The results show that the in-memory database is efficient in analyzing patient cohorts with hierarchical and k-means clustering. On average PAL is faster than R, since R has to perform file-based tasks. When changing the number of dimensions, the database's calculation time remains constant while R scales linearly with the number of dimensions.

Using R is a very fast way to statistically analyze data, but the loading and conversion of data slows down the process. The usage of in-memory technology overcomes this drawback by executing the required algorithms directly within the database layer.

The evaluated clustering algorithms are often used in combination with gene selection algorithms, such as support vector machines, as described in Section 6.2. Therefore, the performance of these algorithms with in-memory technology should be analyzed in future work, since their combination would present a great opportunity for the user to explore patient cohorts using a single platform.

6.7 References

[139] Bell D et al. (2011) Integrated Genomic Analyses of Ovarian Carcinoma. https://tcga-data.nci.nih.gov/docs/publications/ov_2011/. Accessed Sep 23, 2013

[140] Brazma A et al. (2000) Gene expression data analysis. FEBS letters 480(1):17–24

[141] Chen L (2006) Ranking-Based Methods for Gene Selection in Microarray Data. PhD thesis, University of South Florida

[142] Färber F et al. (2012) SAP HANA Database: Data Management for Modern Business Applications. Sigmod Record 40(4):45–51

[143] Ganzer R et al. (2013) Fourteen-year oncological and functional outcomes of high-intensity focused ultrasound in localized prostate cancer. BJU International

[144] Glenn ND (2005) Cohort analysis, vol 5. SAGE Publications, Incorporated

[145] Jiang D, Tang C, Zhang A (2004) Cluster Analysis for Gene Expression Data: A Survey. IEEE Trans on Knowl and Data Eng 16(11):1370–1386

[146] Johnson SC (1967) Hierarchical Clustering Schemes. Psychometrika 2:241–254

[147] Kaatsch P et al. (2012) Krebs in Deutschland 2007/2008. http://www.krebsdaten.de/Krebs/DE/Content/Publikationen/Krebs_in_Deutschland/kid_2012/krebs_in_deutschland_2012.pdf?__blob=publicationFile. Accessed Sep 23, 2013

[148] Kandoth C et al. (2013) Integrated genomic characterization of endometrial carcinoma. https://tcga-data.nci.nih.gov/docs/publications/ucec_2013/. Accessed Sep 23, 2013

[149] Kanungo T et al. (2002) An efficient k-means clustering algorithm: analysis and implementation. Pattern Analysis and Machine Intelligence, IEEE Transactions on 24

[150] Keller G (2011) Mathematik in den Life Sciences. UTB GmbH

[151] MacQueen JB (1967) Some Methods for Classification and Analysis of MultiVariate Observations. In: Cam LML, Neyman J (eds) Proc. of the fifth Berkeley Symposium on Mathematical Statistics and Probability, University of California Press, vol 1, pp 281–297

[152] Muzny DM et al. (2012) Comprehensive molecular characterization of human colon and rectal cancer. https://tcga-data.nci.nih.gov/docs/publications/coadread_2012/. Accessed Sep 23, 2013

[153] Network CGAR (2012) Comprehensive genomic characterization of squamous cell lung cancers. https://tcga-data.nci.nih.gov/docs/publications/lusc_2012/. Accessed Sep 23, 2013

[154] Plattner H (2013) A Course in In-Memory Data Management: The Inner Mechanics of In-Memory Databases. Springer

[155] Poeggel G (2009) Kurzlehrbuch Biologie - , 2nd edn. Georg Thieme Verlag, Stuttgart

[156] SAP AG (2012) SAP HANA Database - SQL Reference Manual. http://help.sap.com/hana/SAP_HANA_SQL_and_System_Views_Reference_en.pdf. Accessed Sep 23, 2013

[157] SAP AG (2013) SAP HANA Predictive Analysis Library (PAL) Reference. http://help.sap.com/hana/SAP_HANA_Predictive_Analysis_Library_PAL_en.pdf. Accessed Sep 23, 2013

[158] Sherlock G (2001) Analysis of Large-scale Gene Expression Data. Briefings in Bioinformatics 2(4):350–362

[159] The R Foundation for Statistical Computing (2006) What is R? http://www.r-project.org/about.html. Accessed Sep 23, 2013

[160] Xu R et al. (2005) Survey of Clustering Algorithms. IEEE Transactions on Neural Networks 16(3):645–678

[161] Yeung KY et al. (2001) Model-based clustering and data transformations for gene expression data. Bioinformatics 17(10):977–987

[162] Zhao W, Ma H, He Q (2009) Parallel K-Means Clustering Based on MapReduce. In: Proceedings of the 1st International Conference on Cloud Computing, Springer-Verlag, Berlin, Heidelberg, pp 674–679

[163] Zheng CH et al. (2008) Gene Expression Data Classification Using Consensus Independent Component Analysis. Genomics, Proteomics & Bioinformatics 6(2):74–82

[164] Zvelebil M, Baum J (2007) Understanding Bioinformatics. Garland Science

Chapter 7
Ad-hoc Analysis of Genetic Pathways

Dominik Müller

Abstract Biological pathways describe different processes and relations within a cell and help to understand the human body. Therefore, they can aid in finding the cause of a genetic disease and thus support treatment decisions. However, identifying pathways affected by mutations based on their internal connections is a complex task. Today, most pathway databases offer only a single keyword search to find pathways. Only a small subset of the databases offer a more complex analysis, such as the ConsensusPathDB and hiPathDB, use an approach based on the relationships between genes. In this contribution, I propose a prototype for analyzing pathways based on their internal topology and relations. Over the course of several months, I aggregated the data of multiple pathway databases. Using in-memory database technology, the prototype traverses the underlying graph of these data to find affected pathways based on a set of genes. The possibility to traverse the pathway graph on the fly might help to find new relationships between diseases and pathways.

7.1 Introduction

A cell contains many biochemical molecules. As an example, an important class of molecules are proteins [213, Chap. 2]. They are responsible for the cellular structure, catalyzing chemical reactions, and regulating gene activity. On the other hand, genes encode amino acids, which form proteins. Thus, there is a complex network within a cell.

Pathways provide an integrated picture of these processes. They give deeper insights into how a cell works by showing the connections and interactions between different components, such as genes, proteins and biochemical reactions. Changes to these procedures, as an effect of a mutation, can lead to diseases. However, often there is not just one single mutation responsible for a disease but the interaction of multiple changes [198, 212]. Thus, studying only individual parts does not provide a complete picture.

The creation of a pathway is a sophisticated process [199]. Often the collected knowledge is based solely on literature. Therefore, related work is reviewed to judge the scientific worth. To create a pathway map, the collected knowledge is broken down into the underlying relationships and pieces of information. The last step is usually another review by other scientists. Today, there are tools that help to find pathways, e.g. GENIES, a natural-language processing system for extraction of molecular pathways [177]. The process ensures that pathways provide precise information, but it is still driven by humans, as a lot of individual research and manual verification needs to be done.

With the development of high-throughput genome sequencing, new possibilities arise for the treatment of genetic diseases. As Fernald et al. note: "We are on the verge of the genomic era: doctors and patients will have access to genetic data to customize medical treatment [176]."

The analysis of pathways leads to a deeper understanding of the connection between different processes in the human body. Thus, it allows us to draw new conclusions and gain novel knowledge about the human genome. It was possible to find new connections between diseases, for example deafness and cardiomyopathy, with the help of pathway analysis [178, 178]. As Ramanan et al. explain, "[...] pathway mechanisms are natural sources for developing strategies to diagnose, treat and prevent complex diseases [202]." However, the growing number of pathways makes it hard to immediately find pathways affected by mutations.

In this work, I investigate a method to analyze pathways based on their topology information. Thereby, I use new in-memory technology.

The remainder of my work is structured as follows. In Section 7.2, I will outline the current field of pathway databases and pathway analysis and explain the technical background of a graph and how to store it. My implementation of a unified pathway database is described in Section 7.3. In the same section, I explain how the underlying dataset is traversed to find relevant pathways. I elucidate the user interface with the help of a use case in Section 7.4. In Section 7.5, I explain the benchmark that has been carried out to evaluate the prototype. The results of the benchmark are then discussed in Section 7.6. I will finally conclude my contribution in Section 7.7 and give a brief outlook of future possibilities.

7.2 Related Work

In the following, I will give an overview of existing pathway databases and analysis tools that are relevant in the context of my work. Figure 7.1 shows the structure of a pathway by showing part of the colorectal pathway from the Kyoto Encyclopedia of Genes and Genomes (KEGG) [184, 185]. It displays the different relations between genes, such as activation, inhibition, and phosphorylation, for a specific context. Different effects of cell correlations are shown, as well as other related pathways. Pathways illustrate the biological dependencies within a cell. Thus, they

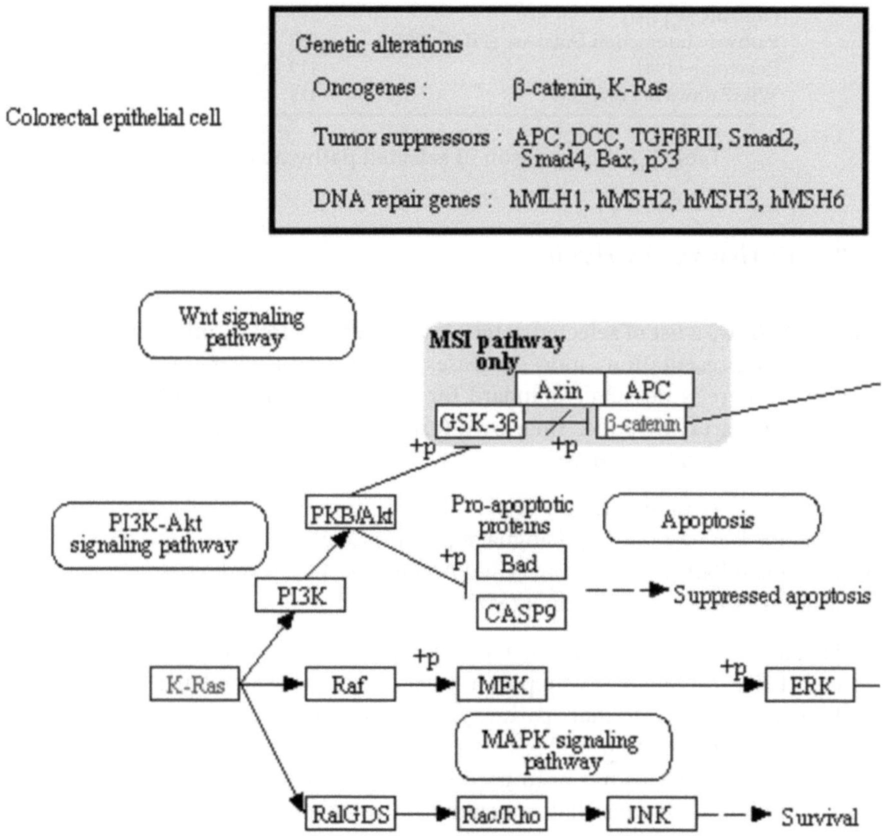

Fig. 7.1: Excerpt from the colorectal cancer pathway taken from KEGG [186].

help us to understand the biological and chemical processes within a cell. The incentive of understanding the human body has resulted in a huge field of curated pathway and tools for analysis. There are already over 300 data resources related to pathways or molecular interaction [165]. In the following sections, I will give a broad overview of the present field and only explain the most relevant parts in more detail.

Data Source	#Pathways	Data Format
BioCarta [197]	354	n/a
KEGG [184, 185]	270	KGML
HumanCyc [205]	297	SBML, BioPAX
Pathway Interaction Database (PID) [208]	137	PID XML, BioPAX
Reactome [194]	1419	MIF, SBML, BioPAX
WikiPathways [187]	415	BioPAX

Table 7.1: Comparison of selected pathway databases.

7.2.1 Pathway Analysis

Table 7.1 shows a list of selected databases that offer pathway information. Besides the online representations, most databases provide free academic access to the data. However, there is no single standard for pathway exchange, and many sources, such as BioCarta, introduce their own forma [190]. Nevertheless, there are some overarching formats, such as:

- Simple Interaction Format (SIF) [172],
- Systems Biology Markup Language (SBML) [182],
- Biological Pathway eXchange (BioPAX) format [173], and
- Molecular Interaction Format (MIF) [188].

Strömbäck and Lambrix give a more comprehensive comparison of the latter three [210]. Most databases support at least one of these standards.

Ramanan et al. state that "pathway analysis represents a potentially powerful and biologically oriented bridge between genotypes and phenotypes [202]." The term pathway analysis has been used in a broader context, while all approaches aim at finding pathways that are significantly affected by a list of genes [180]. For a detailed overview of existing solutions and open challenges, Khatri et al. are referenced [189]. Past approaches mostly fall within one of the categories of Over-Representation Analysis (ORA) or Functional Class Scoring (FCS).

In ORA, a list of genes showing significant changes in expression is compared to a reference set, for example a pathway. The part of genes found in the reference set among the set of genes, which are differentially expressed, is statistically evaluated.

In FCS, a gene-level statistic is calculated for all input genes. For each pathway, these statistics are aggregated resulting in a pathway-level statistic. Based on this statistic a significance value is computed expressing how much the pathway is affected by the set of genes.

In recent years, a new category was established called Pathway Topology Based Approach (PTBA). These approaches use the topology information and connections to calculate the relevance of each pathway. Thus, some of the limitations of ORA and FCS are overcome. For example, due to improved understanding of biochemical processes, pathways will be refined and modified [174]. However, ORA and FCS are not considering the topology of a pathway and thus the calculated value would not change. In the remainder of this work, I refer to pathway analysis as ranking pathways based on their relevance for a specific list of genes.

7.2.2 Existing Analysis Possibilities

There is a wide variety of platforms to analyze and search pathways. All of the databases from Table 7.1 offer at least a single key word search to find specific pathways. In the following sections, I will explain a selection of tools that focus on analyzing pathways rather than creation and curation.

Consensus Pathway Database

The Consensus Pathway Database (ConsensusPathDB) is one of the biggest integrated databases, combining knowledge of other resources [183]. In its current release 26, it integrates 31 other data sources with a total amount of 4,873 pathways, 155,205 unique physical entities and 400,368 unique interactions. Different services are offered to analyze gene sets, like ORA, enrichment analysis, and induced network modules. Nevertheless, only the last one is a PTBA and considers the interactions between genes by finding connections among them, using a list of input genes.

Human-Integrated Pathway Database

The Human-integrated Pathway Database (hiPathDB) is based on the databases PID, Reactome, BioCarta and KEGG [211]. A super pathway has been created from all the resources, which can be queried. Visualization is strongly integrated into hiPathDB for exploring pathways and connections between pathways.

Pathway Express

Pathway Express is a web-based tool for pathway analysis [174]. The tool statistically evaluates, which pathway is significantly impacted from a set of differentially expressed genes. In addition to ORA or FCS, further biological factors are used, such as the position of the genes on the pathway, the topology, the type of interaction, and the magnitude of expression change.

Pathway Projector

Using the KEGG Atlas and Google Maps API, the main approach of this integrated database is a scalable map of the metabolic pathway [192]. It enables physicians to overlay their own data with the map and find data using a keyword search function. The approach focuses on an interactive exploration of the interactions.

7.2.3 Storing a Graph

Big data problems involving complex interconnected data have become important in science in recent years [195]. One example is the analysis of pathways, which contain the biological dependencies within cells. They are commonly represented as a graph. A graph is a data structure composed of vertices and edges [204]. Figure 7.2 shows a small part of the colorectal cancer pathway from KEGG. Each of the genes is represented by a vertex; the relation between two genes by an edge. The relation has an orientation illustrated by the arrow. Additional information is attached to each vertex and edge.

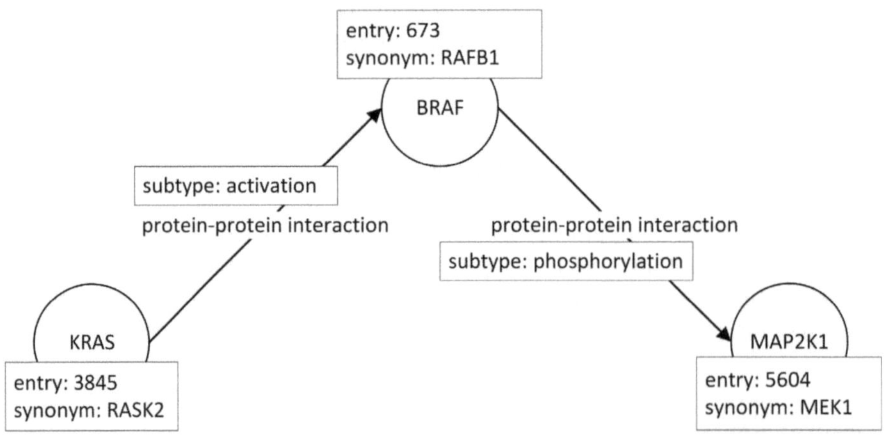

Fig. 7.2: A property graph showing part of a colorectal cancer pathway from KEGG [186]. A circle represents a vertex, whereas an arrow between two vertices indicates edges. Attributes are attached to nodes and vertices in squares.

Mathematically, a graph is defined as $G = (V, E)$ where V is a set of vertices and E is a set of edges. When the edges have an orientation, a graph is called a directed graph. In a directed graph, edges are ordered pairs of vertices or $E = (u, v) \mid u, v \in V$. The following graph types are important for pathways:

- **Vertex-labeled**: vertices can have an identifier,

- **Edge-labeled**: edges can have an identifier for the corresponding relation,
- **Attributed**: additional metadata can be added to edges and vertices,
- **Multi-graph**: there can be more than one edge between the same two vertices,
- **Directed**: edges have an orientation, and
- **Hypergraph**: edges can connect any number of vertices.

As shown in Figure 7.2, these types can be combined with one another. A property graph is a directed, labeled, attributed multi-graph. Since it implicitly supports most other graph types, it is supported by many graph systems. Furthermore, most pathway databases use a property graph as underlying data model.

Graphs have been studied for a long time. The first paper about graph theory is dated back to 1735 when Leonhard Euler defined a path in a graph to be a sequence of vertices and edges, $v_0, e_1, v_1, e_3, ..., v_{n-1}, e_n, v_n$ in which each edge e_i connects the vertices v_{i-1} and v_i $1 \leq i \leq n$ [167, Chap. 1].

Graph theory has studied different properties of a graph, such as connectivity or shortest path. For the pathway analysis, it is important to check if two genes are connected in a pathway by any number of relations. This means that there is a path between the two vertices v_0 and v_n, which represent the genes.

7.2.4 NoSQL

There are different ways to store a graph and therefore pathways. For a long time, the relational database model has been the primary model of database technology. However, more and more database systems based on individual storage approaches have emerged, e.g. NoSQL [171, 193, 209]. A brief introduction of different database types is given in the following.

- **Document database**: The atomic entity is a so-called document. Collections of these documents are used to store and organize the information. The data has a standard format, for example eXtensible Markup Language (XML) or JavaScript Object Notation (JSON). Document databases scale well horizontally.
- **Key-value store**: A map or dictionary is the underlying data model storing values indexed by a key. Their main advantage is that key-value stores scale good for large amounts of data over a number of machines.
- **Column store**: The data is stored in contrast to relational databases by columns instead of rows. Often, this is advantageous for analytical purposes.
- **Graph database**: The data model is a graph with nodes, edges and properties, which is good for interconnected data.

The comparison of the different types in Table 7.2 shows that there are specific advantages and disadvantages for each type. Any of them could be used to store and process a graph. However, a graph database is constructed for complex and flexible data. Also, the underlying model of a graph provides the functionality of graph theory, which is lacking in other databases. Elements of a graph database have direct references to their adjacent elements. Thus, a graph database has an

	Key-value Stores	Column Stores	Document Stores	Graph Databases	Relational Databases
Performance	H	H	H	V	V
Scalability	H	H	V (H)	V	V
Flexibility	H	M	H	H	L
Complexity	N	L	H	H	M
Functionality	V (N)	L	V (L)	GT	RA

Table 7.2: Comparison of database models (H = High, L = Low, V = Variable, M = Moderate, N = None, GT = Graph theory, RA = Relational algebra) [181].

advantage in traversing graphs [204]. Traditional databases would take a great deal longer with this operation, due to the recursive nature of traversing [170].

7.2.5 In-memory Database Technology

The development of computer systems with multiple cores and a lot of main memory has allowed the rise of in-memory databases technology. These databases use the main memory for storage and therefore save read and write time. Consequently, these databases are faster than traditional ones. Additionally, the combination of analytical and transactional processing is enabled by using both row and column-oriented storage [200]. There are quite a few advantages to in-memory database technology that can be useful for and can be applied to analyzing pathways.

As described in Section 1.3, in the past, increasing the processor's clock speed has also increased the processing power. In today's systems, however, this has changed to an increase in number of cores per Central Processing Unit (CPU). Therefore, parallelization is an important part in today's multi-core architecture for scaling [200]. However, parallelization has inherent limits. A task can only be parallelized to a certain degree, dependent on the largest subtask that needs to be run sequentially. Since each pathway is an independent sub graph, they can be analyzed independently. Thus, parallelization is important for an ad-hoc solution.

The data should reside in the main memory to use all advantages of in-memory technology. Thus, in order to store as much data in main memory as possible, the data should be as small as possible. This can be achieved by using lightweight compression, such as dictionary encoding [200]. Additionally, compression helps to speed up processing time since it reduces the data volume that is transferred between main memory and CPU. Even though pathways are complex networks, there are redundant parts like gene names that can be compressed using this technique.

Additionally, extending the in-memory database with a graph engine enables the following graph specific features:

- Relationships are on the same level as entities,
- Increased schema flexibility, and

```
Call WIPE('
  USE WORKSPACE uri:pathway;
  $kras = { uri:KRAS };
  $map2k1 = $kras->uri:directinteraction(2,2);
  RESULT uri:result FROM $pik3ca;
')
```

Listing 7.1: WIPE source code example

- No strict separation between data and metadata.

SAP's HANA as state-of-the-art in-memory database offers such a graph engine called Active Information Store (AIS) [168, 175, 206]. AIS offers the possibility to represent graph-structured data in a more natural way. The underlying data model is a property graph where info items are vertices and associations are edges. There is one separate table for both types storing all the data and metadata.

AIS has its own query language called Weakly-structured Information Processing and Exploration (WIPE). Using this graph specific language, typical graph operations, such as single or multi-step traversal, are supported. This enables the efficient execution of operations that would be difficult to express in SQL [206]. An example statement is shown below in Listing 7.1, illustrating how the graph from Figure 7.2 is traversed.

First, the workspace that should be used is defined. A workspace is a schema with the tables `INFOITEMS` and `ASSOCIATIONS`. A node in the graph can be reached by its Uniform Resource Identifier (URI), as shown in the example, the node for the gene KRAS is saved in the `kras` variable. The graph is then traversed exactly two steps to the gene `MAP2K1` and again saved in the corresponding variable. In the end, a result table is created with the reached node in the `map2k1` variable.

7.3 Creating an Integrated Pathway Database

Utilizing the power of the in-memory database graph engine, I constructed a prototype for pathway analysis as defined in Section 7.2.1. The following sections describe the different steps necessary to do this.

Collecting Data

As described in Section 7.2, there are many different sources offering pathway knowledge. As the basis of my prototype, I used the aggregated knowledge of the three databases KEGG, BioCarta, and Reactome. Table 7.3 gives a detailed overview of the collected knowledge. The data from KEGG comprises 263 pathways, which

Name	Pathways	Vertices	Edges
BioCarta	254	12,010	49,155
KEGG	263	32,784	90,682
Reactome	896	27,180	131,653
Own additions		7,182	45,845
Total	1,413	79,156	317,335

Table 7.3: Integrated pathway databases.

are not freely available in any of the formats mentioned in Section 7.2.1, but as KEGG Markup Language (KGML) files. For BioCarta as well as Reactome, the BioPAX download from PID was used. This includes 254 pathways for BioCarta and 896 for Reactome.

Using Python, both formats were parsed to collect the contained information. In total, the gathered data consists of 1,413 collected pathways with 7,182 distinct genes. I enhanced the existing data by adding identifiers for the genes and the edges to the representing pathway entries. This will be explained in more detail in the next sections.

Creating a Unified Pathway Database

As Yu et al. explain, "[...] pathway integration is still a non-trivial task since each pathway database was constructed based on its own pathway model [211]." My data structure is based on a property graph to enable a topology-based analysis as described in Section 7.2. Additionally, a graph has the necessary flexibility for combining the collected data.

The main types of implemented nodes are `pathway`, `identifier` and `entry`. These nodes can be connected with the main types of edges, which are `represents`, `relatedTo` and `entryOf`. Thereby, `represents` connects an `entry` to an `identifier`, `relatedTo` connects two `entries` and `entryOf` connects an `entry` to a `pathway`. Figure 7.3 shows an example to clarify the relationships between these types of nodes and edges.

These abstract nodes and relations have attributes to store the gathered additional information. These attributes include, among others, the specific types of an `entry`, such as gene or compound, or a `relatedTo` relation, such as activation or inhibition. The full list of attributes that can be attached to nodes and edges is shown in Table 7.4. As an example, entries can range from genes and proteins to compounds, and the specific type is noted in the `entryType` attribute.

Since pathways can be hypergraphs, some parts need to be resolved, for example a chemical reaction that is catalyzed by a protein. This construct can be displayed

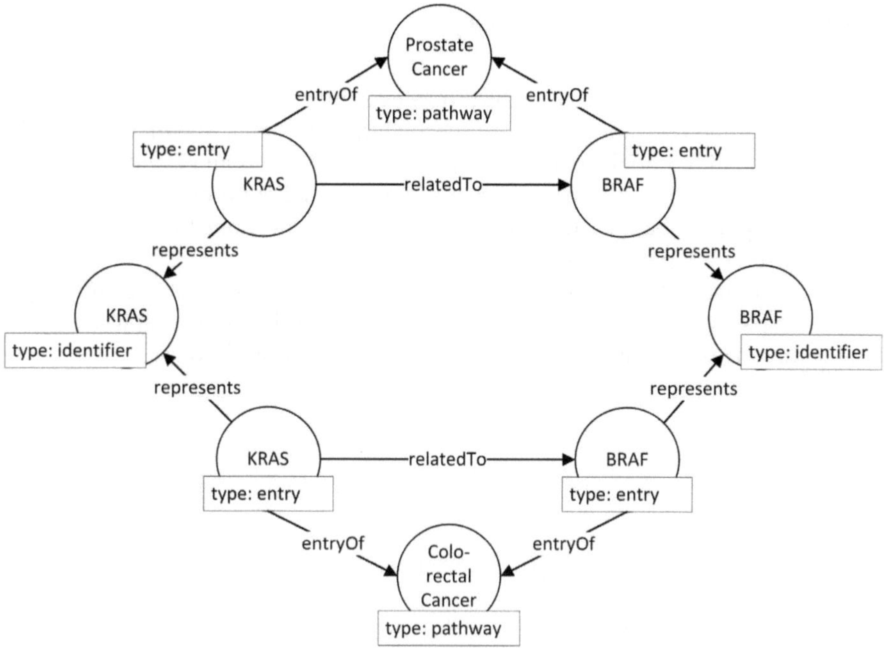

Fig. 7.3: A graph showing the main node and edge types in an example. The genes KRAS and BRAF have one identifier each. This identifier is referenced by an entry that represents that gene in a specific pathway.

name
entryType
pathwayAttribute
image
info
link
reaction

(a) Node attributes

name
relType
pathwayAttribute
value

(b) Edge attributes

Table 7.4: List of attributes for nodes and edges of the created graph.

in a property graph by using an additional node that represents the reaction with a relation from the catalyst to the reaction.

The last step is the unification of identifiers. While one source of data is consistent in nomination, using multiple datasets, the identification needs to be adjusted. Unified identifiers connect pathways as shown in Figure 7.3. This has been achieved by using a dictionary based on the aliases from gene cards [207]. The identifiers are Hugo Gene Symbols [179].

Analyzing Enhanced Pathways

Based on the specific internal architecture explained in Section 7.3, I created an algorithm to find relevant pathways. The analysis is twofold. In both cases, the input is a list of differentially expressed genes. The first part is an FCS-like approach, which finds the number of included genes in the input list for every pathway. Secondly, as topology based criterion, the number of connections between the input genes within each pathway is calculated.

As described in Section 7.2, a pathway can be represented as a graph. In the following paragraph, I explain the parts of the analysis based on graph theory.

I define the set of pathways Γ and the set of genes M as follows.

$$\Gamma = \{G_1, G_2, ..., G_n\},\ G_i = (V_i, E_i), \tag{7.1}$$

$$M \subseteq \bigcup_{i=1}^{n} V_i. \tag{7.2}$$

The path between two vertices v_a, v_b is defined by:

$$\rho(v_a, v_b). \tag{7.3}$$

Based on these definitions, the analysis algorithm needs to calculate the following two parts:

$$p_i = |G_i \cap M|, m_{v_i} = |V_i|, m_{e_i} = |E_i|, \text{ for } 1 \leq i \leq n, \text{ and} \tag{7.4}$$

$$\begin{aligned} p_{e_i} &= \sum_{a=1}^{m_{v_i}} \sum_{b=a+1}^{m_{v_i}} w_{a,b}^{(i)} \text{ for } v_a, v_b \in M, 1 \leq i \leq n, \\ &\text{where } w_{a,b}^{(i)} = \begin{cases} 1 & \text{if } \exists \rho(v_a, v_b), \\ 0 & \text{else.} \end{cases} \end{aligned} \tag{7.5}$$

Equation 7.4 is realized by a lookup in the table followed by an aggregation. The second part shown in Equation 7.5 is done by traversing the underlying graph. Starting from the set of identifier nodes representing the mutated genes, all entries of these genes in every pathway are found. Afterwards, the path is traversed from all those nodes to get the set of reachable nodes. The intersection from the mutated genes and the reached nodes is the set of found connection. This set is grouped by the corresponding pathways and aggregated, resulting in a count of connections for each pathway. The corresponding WIPE statement for two mutated genes is shown below in Listing 7.2.

```
Call WIPE('
  USE WORKSPACE uri:pathway;
  $identifier1 = { uri:KRAS };
  $identifier2 = { uri:MAP2K1 };
  $entries1 = $identifier1<-uri:represents(1,1);
  $entries2 = $identifier2<-uri:represents(1,1);
  RESULT uri:result1
    FROM $entries1->relatedTo(1,*)
    INTERSECTION $entries2;
  RESULT uri:result2
    FROM $entries2->relatedTo(1,*)
    INTERSECTION $entries1;
')
```

Listing 7.2: WIPE example code showing how all connections between two genes are found.

7.4 Application Example

The designed prototype was integrated into the High-performance In-memory Genome (HIG) project, which consists of the following three parts: application layer, platform layer, and data layer, as described in Section 1.5. The architecture is modeled as Fundamental Modeling Concepts (FMC) block diagram in Figure 7.4 [191]. Each layer is adapted to the pathway analysis. The relevant pathway information resides in the data layer, as described in Section 7.3. The platform layer with the integrated graph engine AIS is used for the calculation. Additions to the HIG website have been made in the application layer. These extras are described in the next section.

Use Case and User Interface

I added the following extensions of the HIG platform. I added the User Interface (UI) for manual pathway analysis and integrated the automatic pathway analysis within the results view of the alignment coordinator as described in Section 1.5.1. For illustration purposes of how the system could be used, I assume the following use case.

A physician wants to base the future treatment decision for a patient on the genomic profile of the patient. Thus, the genome has been sequenced. From the list of mutations, the relevant changes have to be found.

Viewing a finished task starts the automatic pathway search. As explained in Section 7.3, the algorithm needs a list of mutated genes to calculate the pathways. The mutations of the corresponding task are used by default. As can be seen in Figure 7.5, the retrieved pathways are shown in a table next to the task results. The list is sorted by the number of connections in the pathway.

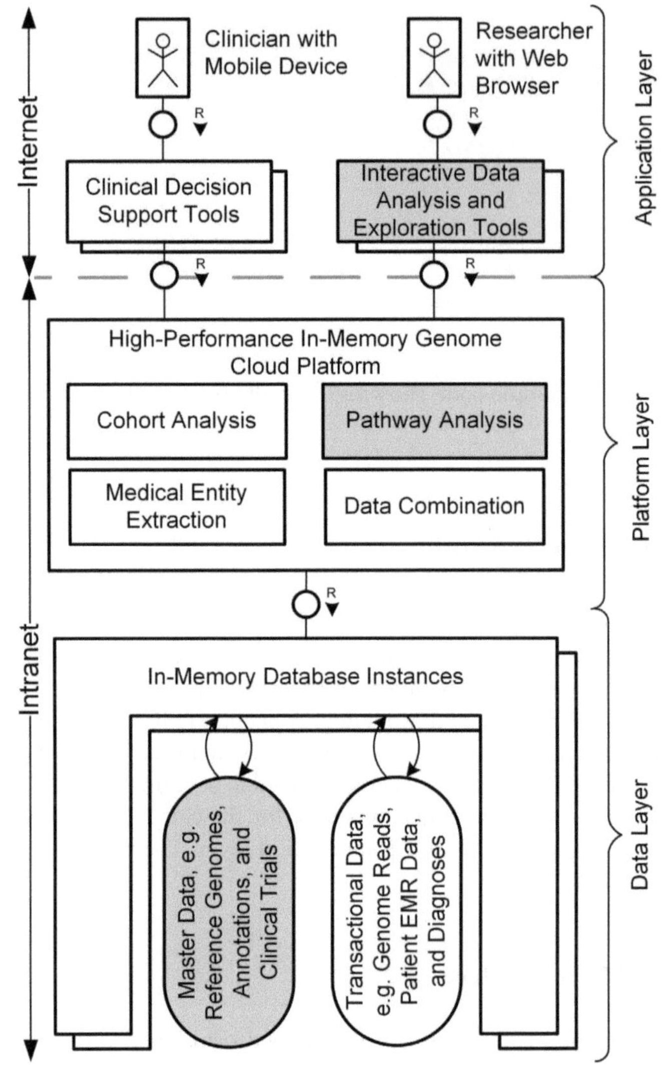

Fig. 7.4: Prototype architecture based on HIG modeled as FMC block diagram.

The ad-hoc search enables the physician to get an overview of the affected pathways without the need to copy genes to another system or tool. Thereby, he can get a deeper understanding of the effects of the mutations and find critical sections that are at risk. Thus, the treatment of the patient can be handled accordingly. This view has the great advantage of combining everything at the same location. There is no need for an upload of a file or copying a list of mutated genes.

Pathway	Database	Connections	Reached	Mutations
Glioma - Homo sapiens (human)	KEGG	9	27	5
Pathways in cancer - Homo sapiens (human)	KEGG	6	58	5
PI3K-Akt signaling pathway - Homo sapiens (human)	KEGG	6	57	5
Melanoma - Homo sapiens (human)	KEGG	6	9	5
Prostate cancer - Homo sapiens (human)	KEGG	4	28	5
Non-small cell lung cancer - Homo sapiens (human)	KEGG	4	18	4
Focal adhesion - Homo sapiens (human)	KEGG	3	51	4
Regulation of actin cytoskeleton - Homo sapiens (human)	KEGG	3	36	3
Colorectal cancer - Homo sapiens (human)	KEGG	3	20	4
Neurotrophin signaling pathway - Homo sapiens (human)	KEGG	3	20	4
Acute myeloid leukemia - Homo sapiens (human)	KEGG	3	11	4
HTLV-I infection - Homo sapiens (human)	KEGG	3	9	5

Fig. 7.5: Automatic pathway search.

On the other hand, in the extra tab, the user can define the list of genes. Figure 7.6 shows how the genes can be chosen with the help of four dropdown fields offering all possible genes. After choosing the requested genes, the pathway analysis can be started. The retrieved pathways are shown in a separate table. Clicking on a pathway can attain additional information such as an image of the pathway. This view enables the exploration of different relations for gene combinations.

7.5 Benchmarks

The benchmarks were carried out to evaluate the viability of the created in-memory prototype in future pathway analyses. The focus lies on the topology-based part of the algorithm, as explained in Section 7.3.

The following impact factors on the execution time of the pathway analysis prototype are investigated with the benchmarks:

1. The number of occurrences of analyzed genes,
2. The number of genes in input set, and
3. The number of integrated pathways.

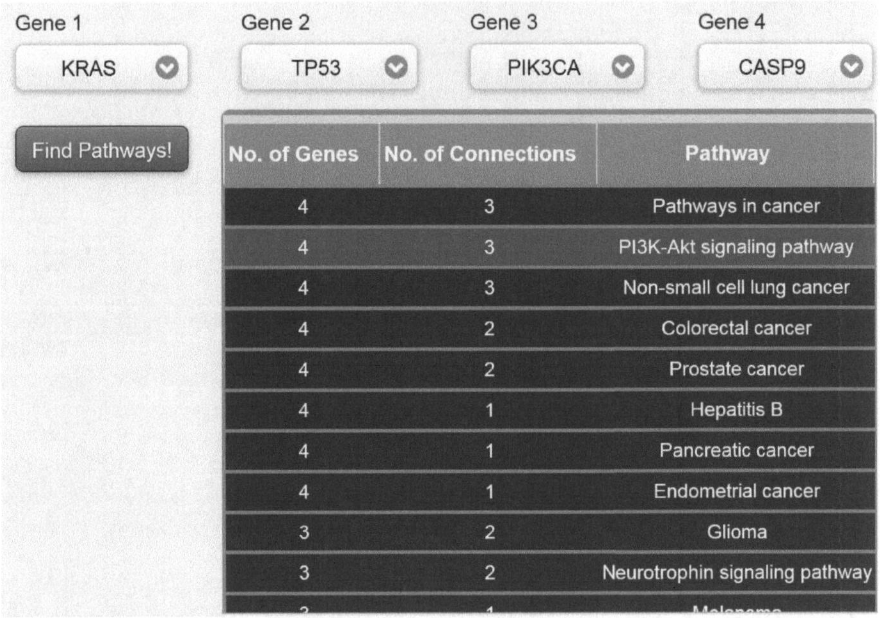

Fig. 7.6: Manual pathway search.

All benchmarks have been done on a system of two identical servers. The servers' specification is shown in Table 7.5. They each had a main memory capacity of 3 TB. SUSE Linux Enterprise Server 11.2 was used as operating system.

Processor	8 x Intel® Xeon® E7-8870 @ 2.40 GHz
Main Memory	64 x 32 GB DDR3 1066 MHz + 64 x 16 GB DDR3 1066 MHz
Operating System	SUSE Linux Enterprise Server 11.2 (Kernel 3.0.13-0.27-default)

Table 7.5: Configuration of benchmark system.

A Python script run by Python 2.64 executed each benchmark and the module `timeit` was used to measure the time. Due to the operating system, `timeit` measured the time with a higher precision then one microsecond [201]. The wall clock time was measured and as a consequence, it has to be considered that other processes may interfere with the timing. Each of the settings was run multiple times until the standard deviation of the considered confidence interval was less than five percent. However, the tests were run at least 50 times and at most 300 times. The confidence interval was set to 90 percent to remove rogue results, such as temporary latency spikes. Each execution consisted of two statements, one being the

WIPE part for traversing the graph and the other one being the SQL statement to aggregate intermediate results.

N	L	H	Genes
50	10	10	KREMEN2, ST8SIA1, TNFSF10, ATF6
150	30	30	HADHB, BAX, MAPK13, CYP1A1, ATF2
250	49	51	NFKBIA, PLCB1, ITGB1, MYC, KRAS
350	69	71	PRKACB, FOS, PRKACG, FASN, NFKB1
450	85	95	RAC1, MAP2K2, JUN, TP53, RELA
551	100	119	RAF1, GRB2, PIK3CA, RPS27A, MAPK8
647	117	145	HRAS, MAP2K1, AKT1, RAF1, GRB2
765	125	185	MAPK3, MAPK1, HRAS, MAP2K1, AKT1

Table 7.6: Used gene sets for first benchmark (N = Number of occurrences, L = Lowest occurrence, H = Highest occurrence).

As reference for this benchmark, I used the graph database Neo4j Enterprise Edition version 2.0.0 Milestone 3, which I installed on one of the servers [196]. The data had been loaded into Neo4j with the help of Gephi, GraphML and Gremlin [166, 169, 203]. The used query language was Cypher. The initial heap size was set to 8 GB and the maximum to 64 GB. As a result of the different execution times on Neo4j, the conditions were as follows. Each test was executed at least five times and the required standard deviation was 15 percent whereas the maximum number of executions was set to 20.

7.5.1 Occurrences of Analyzed Genes

The first benchmark investigates to what extent the number of occurrences of the chosen genes influences the time of the analysis. As explained in Section 7.3, a gene has an identifier node, which is referenced by each entry that represents the gene. Thus, more occurrences of a gene mean more entries need to be traversed. The used genes and the number of occurrences are shown in Table 7.6. This benchmark was executed on the gathered data and a set of five genes.

7.5.2 Genes in Input Set

The second benchmark aims to determine how the performance develops when the number of analyzed genes increases. For that, I varied the set sizes, i.e. I used five, and from ten to 100 I increased the set in steps of ten. In each pass the first n genes with the most occurrences are used. The gathered data was used for this benchmark setup.

7.5.3 Integrated Pathways

The last benchmark investigates how the proposed prototype scales with an increase of the used dataset. It can give insights into how the execution time of the prototype develops for future scenarios where more pathways are integrated.

Additional data was generated for this benchmark based on the collected real data. The generated data samples are based on the collected real data and have minor changes, such as deletion or interchange of vertices and edges. The exact number of the test sets is shown in Table 7.7. The analyzed genes were MAPK3, MAPK1, HRAS, MAP2K1, and AKT1.

Factor	Pathways	Vertices	Edges
1	1,413	79,203	317,384
2	2,826	144,037	581,797
5	7,065	338,530	1,359,322
10	14,130	662,844	2,609,769
20	28,260	1,311,243	4,964,225
30	42,390	1,959,792	7,166,956
40	56,520	2,608,300	9,266,036
50	70,650	3,256,789	11,291,154
60	84,780	3,905,308	13,262,129
70	98,910	4,553,826	15,199,371
80	113,040	5,202,334	17,116,761
90	127,170	5,850,814	19.038.912
100	141,300	6,499,312	20,960,354

Table 7.7: Comparison of database sizes of the test sets for the third benchmark.

7.6 Results and Discussion

In the following chapter, I will evaluate the benchmarks, which were executed as described in the previous chapter.

7.6.1 Occurrences of Analyzed Genes

Table 7.8 shows the results of the first benchmark. I exemplarily ran some tests with Neo4j. With 50 occurrences, they took two seconds to run; with 250 occurrences, they took twelve seconds, and with 350 occurrences, they took almost 77 seconds.

These results imply that AIS is a better fit for the use case of pathway analysis. A possible explanation is the structure of the used queries. The IMDB looks up

N	Mean [s]	SD [%]
50	0.171	4
150	0.201	6
250	0.232	3
350	0.224	2
450	0.272	3
551	0.274	3
647	0.271	4
756	0.272	3

Table 7.8: Results depending on the number of analyzed genes (N = Number of occurrences, SD = Standard deviation).

each identifier once and creates the set of all intersected genes. An example for the first set of genes is given in Section 7.8.

Furthermore, the used `UNION ALL` statement in Cypher was just released with Neo4j 2.0.0, which is still in development. Thus, it might not be optimized yet. As an example, it is not certain to what extent the sub queries are done in parallel while in AIS different traversal parts are automatically parallelized.

The results verify the hypothesis that the execution time depends on the chosen input genes. This supports the decision of using a fixed set of genes in the other benchmarks. However, Figure 7.7 does not imply a direct linear correlation between time of execution and the number of occurrences of the input genes. Thus, further tests should be conducted to investigate in other impact factors, such as number of reachable nodes or number of connections found.

7.6.2 Genes in Input Set

The number of mutated genes in a patient can vary greatly. This benchmark was executed to find out about the influence of the number of input genes on the execution time of the analysis.

The results in Table 7.9 show that with an increasing number of input genes the execution time grows, which is to be expected. Each gene increases the total number of occurrences, reached nodes and possible connections. Additionally, multiple parts for each gene extend the statement. By offering the possibility of iterating over the set of vertices this could be improved. The results are illustrated in Figure 7.8 where lines are drawn between the measurement results to guide the eye. The growth can be explained by the growing complexity of possible connections between the genes. Finding connections between vertices is a typical graph operation, in my opinion. Thus, offering this as built-in function would allow for internal optimization and therefore further improve the performance.

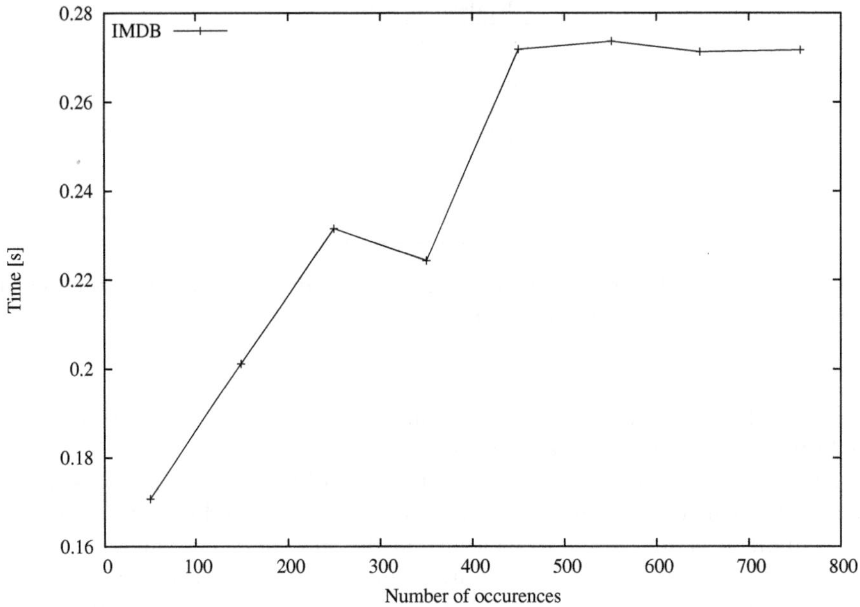

Fig. 7.7: Calculation time depending on the input genes.

#Genes	Mean [s]	SD [%]
5	0.287	6
10	0.459	2
20	0.772	5
30	1.172	9
40	1.503	8
50	1.979	10
60	2.309	3
70	2.837	8
80	3.375	9
90	3.769	8
100	4.405	8

Table 7.9: Results depending on the genes in input set (SD = Standard deviation).

7.6.3 Integrated Pathways

In the future, more pathways will be created and curated. Thus, the number of pathways and the amount of information within each pathway will rise. The main tissue types alone would result in quadrupling the number of pathways. This benchmark was conducted to test how the prototype handles those future scenarios.

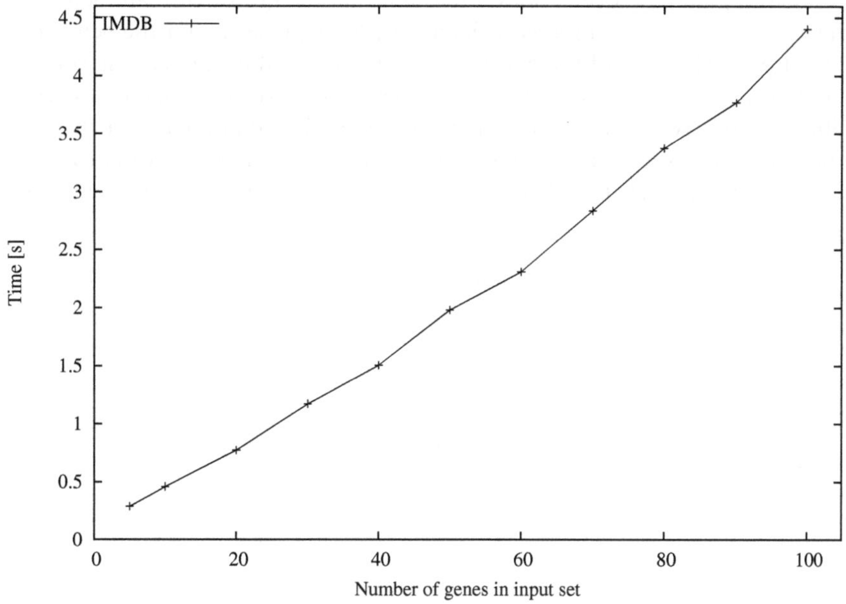

Fig. 7.8: Calculation time depending on the gene set size.

The results are displayed in Table 7.10. They show that even with ten times the amount of pathways the query stays under one second. There is a common

#Pathways	Mean [s]	SD [%]
1,413	0.284	5
2,826	0.325	4
7,065	0.574	5
14,130	0.975	5
28,260	1.494	5
42,390	1.657	5
56,520	2.086	5
70,650	1.925	2
84,780	2.267	4
98,910	2.451	3
113,040	2.668	3
127,170	2.686	3
141,300	2.870	2

Table 7.10: Results depending on the number of integrated pathways (SD = Standard deviation).

trend, which is illustrated in Figure 7.9. The lines between the measurement results are drawn to guide the eye. It shows that the runtime grows with an increase of pathways while the rate of increase decreases. An explanation for the behavior is the good parallelization of the queries. The number of starting genes is constant and only the number of instances of the genes in new pathways increases. These additional pathways can all be searched independently. Additionally, the overhead of calculating the execution plan does not increase significantly with an increase of pathways. This suggests that AIS is a good fit for increasing numbers of pathways.

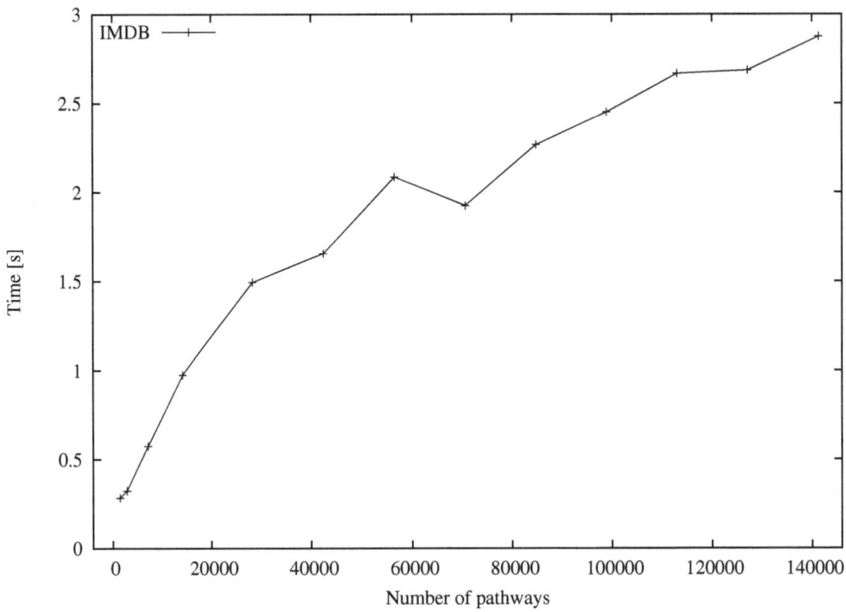

Fig. 7.9: Calculation time depending on pathway size.

7.7 Conclusion and Outlook

I proposed a pathway analysis tool that uses in-memory database technology to find affected pathways from a list of genes based on the underlying topology. I used pathways from multiple pathway databases. To combine this knowledge I unified them into one graph. This representation allows finding connections between genes in a pathway by traversing the graph. Thus, the topology of the pathways is considered unlike in most of today's solutions.

While today a lot of information is omitted, there will be an increasing number of pathways, e.g. by adding cell- and tissue-specific information. The benchmark showed that the designed prototype can handle the increasing amounts of data.

With other parts integrated into in-memory database technology like text search and analytical libraries, new kinds of pathway analysis could be possible, such as statistical analysis of the underlying graph or a combination of the complex knowledge of pathways and text analysis. In my opinion, these kinds of analysis can further strengthen the support of treatment decisions.

By further improving the performance of the graph engine, additional types of analyses would be possible. Enabling the scientist or physicians to interact with the ad-hoc analysis would create a whole new level of experience. Editing a pathway and directly getting the effects and impacts of these changes on the global network of molecules, pathways, and diseases would lead to an even greater understanding of the human body. This could even expand to a dynamic model helping to find new treatment possibilities.

7.8 Appendix

In the following, the relevant queries used during my benchmarks are shown. Listing 7.3 contains the WIPE query, Listings 7.4 and 7.5 show the corresponding queries for SQL and Cypher, respectively.

```
Call WIPE('USE WORKSPACE uri:test1WS;
$id0 = { uri:MAPK3 }; $id1 = { uri:MAPK1 };
$id2 = { uri:HRAS }; $id3 = { uri:MAP2K1 };
$id4 = { uri:AKT1 };
$gene0 = $id0<-uri:represents(1,1);
$gene1 = $id1<-uri:represents(1,1);
$gene2 = $id2<-uri:represents(1,1);
$gene3 = $id3<-uri:represents(1,1);
$gene4 = $id4<-uri:represents(1,1);
$allGenes = { uri:MAPK3, uri:MAPK1,
 uri:HRAS, uri:MAP2K1, uri:AKT1 }<-uri:represents(1,1);
RESULT uri:reached0
FROM ($gene0->uri:relatedTo(1,*)
INTERSECTION ($allGenes DIFFERENCE $gene0));
RESULT uri:reached1
FROM ($gene1->uri:relatedTo(1,*)
INTERSECTION ($allGenes DIFFERENCE $gene1));
RESULT uri:reached2
FROM ($gene2->uri:relatedTo(1,*)
INTERSECTION ($allGenes DIFFERENCE $gene2));
RESULT uri:reached3
FROM ($gene3->uri:relatedTo(1,*)
INTERSECTION ($allGenes DIFFERENCE $gene3));
RESULT uri:reached4
FROM ($gene4->uri:relatedTo(1,*)
INTERSECTION ($allGenes DIFFERENCE $gene4));')
```

Listing 7.3: Example WIPE query as used in benchmark

```
SELECT "pathwayAttribute"
FROM(
    SELECT "pathwayAttribute" FROM
        "test1WS"."reached0" UNION ALL
    SELECT "pathwayAttribute" FROM
        "test1WS"."reached1" UNION ALL
    SELECT "pathwayAttribute" FROM
        "test1WS"."reached2" UNION ALL
    SELECT "pathwayAttribute" FROM
        "test1WS"."reached3" UNION ALL
    SELECT "pathwayAttribute" FROM
        "test1WS"."reached4")
```

Listing 7.4: Example SQL query as used in benchmark

```
Start g0 = node:node_auto_index(label="KREMEN2"),
      g1 = node:node_auto_index(label="ST8SIA1"),
      g2 = node:node_auto_index(label="TNFSF10"),
      g3 = node:node_auto_index(label="ATF6")
MATCH (g0)<-[:represents]-(a)-[:relatedTo*]->(b)
-[:represents]->(c)
WHERE c = g1 OR c = g2 OR c = g3
RETURN b.pathwayAttribute as pathway
UNION ALL
Start g0 = node:node_auto_index(label="KREMEN2"),
      g1 = node:node_auto_index(label="ST8SIA1"),
      g2 = node:node_auto_index(label="TNFSF10"),
      g3 = node:node_auto_index(label="ATF6")
MATCH
(g1)<-[:represents]-(a)-[:relatedTo*]->(b)
-[:represents]->(c)
WHERE c = g0 OR c = g2 OR c = g3
RETURN b.pathwayAttribute as pathway
UNION ALL
Start g0 = node:node_auto_index(label="KREMEN2"),
      g1 = node:node_auto_index(label="ST8SIA1"),
      g2 = node:node_auto_index(label="TNFSF10"),
      g3 = node:node_auto_index(label="ATF6")
MATCH
(g2)<-[:represents]-(a)-[:relatedTo*]->(b)
-[:represents]->(c)
WHERE c = g0 OR c = g1 OR c = g3
RETURN b.pathwayAttribute as pathway
UNION ALL
Start g0 = node:node_auto_index(label="KREMEN2"),
      g1 = node:node_auto_index(label="ST8SIA1"),
      g2 = node:node_auto_index(label="TNFSF10"),
      g3 = node:node_auto_index(label="ATF6")
MATCH
(g3)<-[:represents]-(a)-[:relatedTo*]->(b)
-[:represents]->(c)
WHERE c = g0 OR c = g1 OR c = g2
RETURN b.pathwayAttribute as pathway
```

Listing 7.5: Example Cypher query for Neo4j as used in benchmark

7.9 References

[165] Bader GD, Cary MP, Sander C (2006) Pathguide: A Pathway Resource List. Nucleic Acids Research 34(Suppl 1):D504–D506

[166] Bastian M, Heymann S, Jacomy M (2009) Gephi: An Open Source Software for Exploring and Manipulating Networks. In: ICWSM

[167] Bigg NL et al. (1976) Graph Theory 1736-1936. Oxford University Press on Demand

[168] Bornhövd C et al. (2013) Flexible Information Management, Exploration, and Analysis in SAP HANA. In: Proceedings of the International Conference on Data Technologies and Applications, pp 15–28

[169] Brandes U et al. (2010) Graph Markup Language (GraphML). In: Tamassia R (ed) Handbook of Graph Drawing and Visualization, CRC Press, chap 16, pp 517–541

[170] Buerli M (2012) The Current State of Graph Databases. Technical report, Department of Computer Science, Cal Poly San Luis Obispo

[171] Cattell R (2011) Scalable SQL and NoSQL Data Stores. ACM SIGMOD Record 39(4):12–27

[172] Cytoscape Consortium (2012) Cytoscape User Manual: Network Formats. http://wiki.cytoscape.org/Cytoscape_User_Manual/Network_Formats. Accessed Sep 23, 2013

[173] Demir E et al. (2010) The BioPAX Community Standard for Pathway Data Sharing. Nature Biotechnology 28(9):935–942

[174] Draghici S et al. (2007) A Systems Biology Approach for Pathway Level Analysis. Genome Research 17(10):1537–1545

[175] Färber F et al. (2012) SAP HANA Database: Data Management for Modern Business Applications. Sigmod Record 40(4):45–51

[176] Fernald GH et al. (2011) Bioinformatics Challenges for Personalized Medicine. Bioinformatics Journal 27(13):1741–1748

[177] Friedman C et al. (2001) GENIES: A Natural-language Processing System for the Extraction of Molecular Pathways from Journal Articles. Bioinformatics Journal 17(Suppl 1):74–S82

[178] Goh KI et al. (2007) The Human Disease Network. Proceedings of the National Academy of Sciences 104(21):8685–8690

[179] Gray KA et al. (2013) Genenames.org: The HGNC Resources in 2013. Nucleic Acids Research 41(D1):D545–D552

[180] Green, ML and Karp, PD (2006) The Outcomes of Pathway Database Computations depend on Pathway Ontology. Nucleic Acids Research 34(13):3687–3697

[181] Hecht R, Jablonski S (2011) NoSQL evaluation: A use case oriented survey. In: International Conference on Cloud and Service Computing (CSC), 2011, IEEE, pp 336–341

[182] Hucka M et al. (2010) The Systems Biology Markup Language (SBML): Language Specification for Level 3 Version 1 Core. http://precedings.nature.

com/documents/4959/version/1/files/npre20104959-1.pdf. Accessed Sep 23, 2013
[183] Kamburov A et al. (2013) The ConsensusPathDB interaction database: 2013 update. Nucleic Acids Research 41(D1):D793–D800
[184] Kanehisa M, Goto S (2000) KEGG: Kyoto Encyclopedia of Genes and Genomes. Nucleic Acids Research 28(1):27–30
[185] Kanehisa M et al. (2012) KEGG for Integration and Interpretation of Large-scale Molecular Data Sets. Nucleic Acids Research 40(D1):D109–D114
[186] Kanehisa Laboratories (2012) KEGG PATHWAY: Colorectal Cancer - Reference Pathway. http://www.genome.jp/kegg-bin/show_pathway?org_name=map&mapno=05210. Accessed Sep 23, 2013
[187] Kelder T et al. (2012) WikiPathways: Building Research Communities on Biological Pathways. Nucleic Acids Research 40(D1):D1301–D1307
[188] Kerrien S et al. (2007) Broadening the Horizon: Level 2.5 of the HUPO-PSI Format for Molecular Interactions. BMC biology 5(1):44
[189] Khatri P, Sirota M, Butte AJ (2012) Ten Years of Pathway Analysis: Current Approaches and Outstanding Challenges. PLoS Computational Biology 8(2)
[190] Kirouac DC et al. (2012) Creating and Analyzing Pathway and Protein Interaction Compendia for Modelling Signal Transduction Networks. BMC Systems Biology 6(1):29
[191] Knöpfel A, Gröne B, Tabeling P (2005) Fundamental Modeling Concepts. Wiley, West Sussex UK
[192] Kono N et al. (2009) Pathway Projector: Web-based Zoomable Pathway Browser using KEGG Atlas and Google Maps API. PLoS One 4(11)
[193] Leavitt N (2010) Will NoSQL Databases live up to their Promise? IEEE Computer 43(2):12–14
[194] Matthews L et al. (2009) Reactome Knowledgebase of Human Biological Pathways and Processes. Nucleic Acids Research 37(suppl 1):D619–D622
[195] Miller JJ (2013) Graph Database Applications and Concepts with Neo4. In: Proceedings of the 2013 Southern Association for Information Systems
[196] Neo4J Developers (2012) Neo4J. Graph NoSQL Database
[197] Nishimura D (2001) BioCarta. The Computer Software Journal for Scient 2(3):117–120
[198] Peng G et al. (2009) Gene and Pathway-based Second-wave Analysis of Genome-wide Association Studies. European Journal of Human Genetics 18(1):111–117
[199] Pico AR et al. (2008) WikiPathways: Pathway Editing for the People. PLoS biology 6(7):184
[200] Plattner H (2013) A Course in In-Memory Data Management: The Inner Mechanics of In-Memory Databases. Springer
[201] Python Software Foundation (2013) 26.6. Timeit - Measure Execution Time of Small Code Snippets. http://docs.python.org/2/library/timeit.html. Accessed Sep 23, 2013

[202] Ramanan VK et al. (2012) Pathway Analysis of Genomic Data: Concepts, Methods and Prospects for Future Development. Trends in Genetics 28(7):323–332

[203] Rodriguez MA (2013) Gremlin GitHub. https://github.com/tinkerpop/gremlin/wiki. Accessed Sep 23, 2013

[204] Rodriguez MA, Neubauer P (2010) Constructions from Dots and Lines. Bulletin of the American Society for Information Science and Technology 36(6):35–41

[205] Romero P et al. (2004) Computational Prediction of Human Metabolic Pathways from the Complete Human Genome. Genome Biology 6(1):R2

[206] Rudolf M et al. (2013) The Graph Story of the SAP HANA Database. Proceedings of the 15 GI-Fachtagung Datenbanksysteme für Business, Technologie und Web

[207] Safran M et al. (2010) GeneCards Version 3: The Human Gene Integrator. The Journal of Biological Databases and Curation 2010

[208] Schaefer CF et al. (2009) PID: The Pathway Interaction Database. Nucleic Acids Research 37(suppl 1):D674–D679

[209] Strauch C (2011) NoSQL Databases. http://www.christof-strauch.de/nosqldbs.pdf. Accessed Sep 23, 2013

[210] Strömbäck L, Lambrix P (2005) Representations of Molecular Pathways: An Evaluation of SBML, PSI MI and BioPAX. Bioinformatics Journal 21(24):4401–4407

[211] Yu N et al. (2012) hiPathDB: A Human-integrated Pathway Database with Facile Visualization. Nucleic Acids Research 40(D1):D797–D802

[212] Zhao J et al. (2011) Pathway-based Analysis using reduced Gene Subsets in Genome-wide Association Studies. BMC Bioinformatics 12(1):17

[213] Zvelebil M, Baum J (2007) Understanding Bioinformatics. Garland Science

Chapter 8
Combined Search in Structured and Unstructured Medical Data

David Heller

Abstract Today, structured medical data is often considered apart from its unstructured counterpart. When searching for a specific piece of information either structured sources, e.g. genomic variant lists and electronic medical records, or unstructured sources, e.g. medical papers, research documentations and trial descriptions, are examined. However, structured data, such as a patient's genomic data, can be valuable in searching unstructured data like clinical trial proposals in order to find apposite information for the patient. Consequently, today's separation of both source types impedes any insights into coherencies between them. In this contribution, I propose utilizing in-memory databases to combine results from search in structured as well as in unstructured medical data and introduce a research prototype for a clinical trial search tool. The prototype suggests matching clinical trials based on a patient's genome and benefits from the analytical performance of the in-memory database. Furthermore, I investigate how an increasing amount of medical input data affects the performance of the prototype.

8.1 Introduction

The amount of medical data available in online databases is growing quickly, such as in PubMed, which already comprises 22 million biomedical citations with currently over 700,000 citations added each year [230, 249]. A lot of medical information is freely accessible but scattered across several databases that list e.g. genes, proteins, clinical trials, and genetic pathways [252, 243, 223, 224]. This immense and growing amount of medical knowledge plays a crucial role for scientific research and practice but also poses several problems.

One of the greatest challenges remains searching for a specific required piece of information. Reasons for this are the enormous volume and complexity of medical data as well as the heterogeneity of databases, formats and structures [217]. Making the task even more difficult, a significant amount of today's medical data is encoded in the form of unstructured natural language [228]. Scientific publications and patents, medical reports, as well as comments, keywords, or descriptions in database records use natural language to transmit and exchange information [228]. While I consider this unstructured data a substantial part of the world's medical knowledge, its comprehension, analysis and searching is more difficult than for structured data, such as experiment results or genomic variant data.

Much research in this field has been done to extract information either from structured or unstructured medical data. However, the insights that can be gained by combining results from both sources are manifold. A physician could receive information contained in scientific publications that perfectly match his patient's genome. The suggestion of matching clinical trials could hold the key to possibly prolonging the patient's life by providing access to innovative medical products. These are some examples of how associating structured medical data with search results from unstructured natural language texts can be beneficial to physicians and patients.

In-memory database (IMDB) technology enables the highly accelerated analysis of large amounts of data by keeping primary data permanently in memory [231]. Additionally, it allows efficient text mining and extraction of semantic entities by implementing a text engine [214].

In the given work, I present my findings of utilizing IMDBs for the combined search in structured and unstructured medical data. I configured an IMDB to extract biomedical entities from natural language texts and developed a research prototype for a clinical trial search tool that suggests matching clinical trials based on a patient's genome and metadata. Figure 8.1 depicts the three-layer-architecture of the prototype with the IMDB in the center modeled as a Fundamental Modeling Concepts (FMC) block diagram [225].

The remainder of this contribution is structured as follows: In Section 8.2 related research in the field is discussed. Section 8.3 characterizes IMDB text analysis features and how dictionaries and extraction rules can customize them. The research prototype including its required IMDB configurations are explored in Section 8.4. Section 8.5 describes benchmarks of the research prototype and their results, followed by the discussion of the results in Section 8.6. The work is concluded with an outlook in Section 8.7.

8.2 Related Work

This work is embedded into several research fields. Firstly, it strongly depends on algorithms to process information contained in unstructured human language texts. The general research areas that are relevant here are Natural Language Processing (NLP) and Information Extraction. NLP attempts to derive meaning from human language texts whereas Information Extraction goes a step further and deals with the extraction of information contained in unstructured or semi-structured data – often by means of NLP. Important basic NLP tasks include seg-

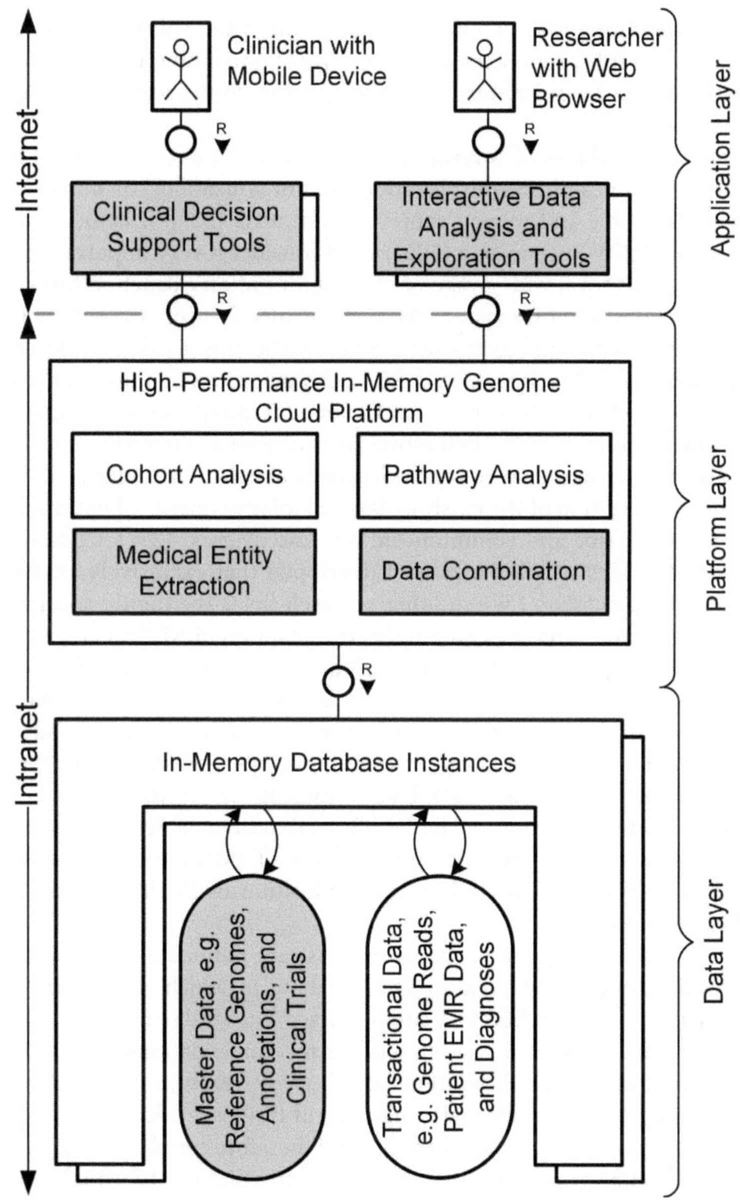

Fig. 8.1: System architecture of my research prototype: Unstructured and structured data is combined and analyzed by the IMDB. The results are accessible through the website and the tablet application.

mentation, stemming and tagging. Segmentation is the splitting of text into its parts like paragraphs, sentences and words [236]. Stemming is a term coined for the identification of word stems. And tagging refers to the part of speech identification of words.

An important subtask of information extraction is the recognition of information units like names of persons, locations and organizations, or numeric expressions including time and money expressions in texts [229]. It is most commonly called Named Entity Recognition (NER) and I consider it very important because it provides the basis for a wide range of other information extraction tasks. A number of approaches to NER exist. Early systems mainly used algorithms based on manually created rules and dictionaries whereas the most modern systems employ machine learning techniques [229]. NER is also incorporated into IMDBs among a range of other text analysis features [214]. For NER IMDBs resort to a combination of predefined dictionaries and extraction rules as well as linguistic models [234].

NLP has a great significance for the biomedical field. Krallinger et al. state that "a considerable fraction of the existing data in biology consists of natural language texts used to describe and communicate new discoveries" [227]. Consequently, a range of text-analysis tools have been developed that exclusively deal with the biomedical domain [226]. Like in other research fields the highly specialized terminology, e.g. medical terms, complicates the adaption of NLP tools developed for generic texts [228]. The constant formations of new terms and abbreviations as well as ambiguity or polysemy are only a few of the challenges. Much research has been done in this area and particularly regarding the extraction and tagging of biomedical entity names like genes, proteins, DNA, RNA and cells from natural language texts [215, 239, 241]. Due to their flexibility this biomedical entity extraction can also be performed by IMDBs as described in Section 8.3.

My work builds on many different aspects of IMDB technology that were explored in recent years, including the application of IMDBs for medical purposes [219, 220, 231, 238].

Clinical Decision Support Systems (CDSSs) are defined as "any software designed to directly aid in clinical decision making in which characteristics of individual patients are matched to a computerized knowledge base for the purpose of generating patient-specific assessments or recommendations that are then presented to clinicians for consideration [222]." My research prototype described in Section 8.4 suggests possibly matching clinical trials specifically for a given patient based on the patient's characteristics and therefore could be incorporated into a CDSS. Sittig et al. emphasize that "there is a pressing need for high-quality, effective means of designing, developing, presenting, implementing, evaluating, and maintaining all types of clinical decision support capabilities for clinicians, patients and consumers [240]." They furthermore identified their top 10 grand challenges in clinical decision support including "3. Prioritize and filter recommendations to the user" and "5. Use free-text information to drive clinical decision support [240]." In this contribution, I address both of these selected challenges by filtering and prioritizing information derived from unstructured natural language text.

```
CREATE FULLTEXT INDEX <index_name>
  ON <table_name>(<text_column_name>)
  CONFIGURATION '<configuration_name>' TEXT ANALYSIS ON
```

Listing 8.1: SQL statement creating a fulltext index

Finally, my work extends the High-performance In-memory Genome (HIG) platform as described in Section 1.5. The HIG platform builds IMDB technology to process and analyze genome data as a holistic process in the course of personalized medicine [237]. My research prototype builds on the HIG system architecture.

8.3 In-memory Database Text Analysis Features

For the extraction of information from unstructured natural texts, IMDBs embed a text analysis engine, which offers text indexing and a wide range of text analysis features [214]. It of the linguistic analysis and the entity extraction component [236].

The linguistic analysis offers NLP capabilities including segmentation, stemming and tagging, whereas the entity extraction part offers NER functionality that discovers entities in the text and identifies their type.

In order to analyze text contained in a database table a fulltext index on the text column has to be created using the SQL syntax from Listing 8.1 [234].

Once the index is created, the text analysis is run in the background. The results are written into an additional result table in the same database schema called `$TA_<index_name>`. The result table contains one row for each token from linguistic analysis or each entity from entity extraction respectively. The rows contain all key columns of the source table to allow linking them to it: the specific term or entity, its linguistic or entity type and additional information about its normalized form and location in the text [234].

The text analysis engine can be customized with dictionaries and extraction rules [235]. In this context, dictionaries are lists of entities with an assigned entity type that enable the database to recognize the listed entities in unstructured text.

A dictionary contains one or more entity types, each of which contains any number of entities [235]. Each entity in turn contains one standard form name and any number of variant names. The dictionary has to be in a specified XML format like shown in Listing 8.2, which defines the entity type `TREE` containing the entity `Lime` with two variant names [235].

Extraction rules, in contrast, define the entities of a specific type using a formal syntax, i.e. the Custom Grouper User Language (CGUL). CGUL is a pattern-based language that enables "pattern matching using character or token-based regular expressions combined with linguistic attributes to define custom entity types" [235]. CGUL is based on tokens detected by the linguistic analysis. It allows

```xml
<?xml version="1.0" encoding="windows-1252"?>
<dictionary>
  <entity_category name="TREE">
    <entity_name standard_form="Lime">
      <variant name="Linden" />
      <variant name="Basswood" />
    </entity_name>
  </entity_category>
</dictionary>
```

Listing 8.2: Example for a dictionary in XML format

```
#group BodyPart: <head> | <foot> | <eye>
```

Listing 8.3: Example for a CGUL group definition

to formulate patterns that match tokens by using a literal string, like `<tree>`, a regular expression, such as `<car.*>`, a word stem, like `<STEM:fly>`, or a word's part-of-speech, such as `<POS:Nn>` for a noun. To define a custom entity type in CGUL the `#group` directive as shown in Listing 8.3 is used. With this directive all occurrences of `head`, `foot` and `eye` will appear in the text analysis result table with the entity type `BodyPart`.

Embedding dictionaries and extraction rules requires the following steps:

1. Creating a file, formatted according to dictionary or extraction rule syntax,
2. Running the dictionary or rule compiler on this file, and
3. Including the compiled dictionary or rule in a database configuration file and referencing this configuration file when creating the fulltext index.

8.4 Application Example

The term clinical trial refers to "research using human volunteers that is intended to add to medical knowledge" [247]. Clinical trials are a vital means of medical research and a crucial step on a new medical product's way to official approval [244]. Many clinical trials are registered in online databases and can thus be analyzed and searched. One of the most comprehensive clinical trial databases is clinicaltrials.gov [245]. As of June 2013, it lists more than 147,000 trials with approximately 400 new records added to the database weekly [254].

The clinicaltrials.gov database can be searched using a basic keyword search or using advanced search fields, such as Conditions, Locations, Gender or Age Group [246]. While this functionality is helpful in many cases it requires the user to supply the right keywords. A combination of the unstructured trial descriptions from clinicaltrials.gov with structured patient data, such as genome sequences, however, can overcome this constraint. It ensures that all available data about a patient is utilized to find a trial matching him. Therefore, I developed a research prototype that enables physicians to find clinical trials matching a specific patient. This prototype is about to be described in this section.

8.4.1 Customized Biomedical Dictionaries

Biomedical dictionaries enable IMDBs to identify biomedical entities in natural language texts and therefore provide the basis of applications for real-time information extraction and analysis from such texts.

A considerable part of today's medical publications deals with genetic correlations [216]. Therefore, the fast and reliable identification and normalization of gene names is very important, which led me to assemble a customized gene dictionary. The identification of pharmaceutical ingredients and drug products in texts is another important task as many publications deal with discoveries made about pharmaceutical drugs and their effects. Therefore, I assembled a dictionary holding pharmaceutical ingredients.

The decision in favor of pharmaceutical ingredients versus drug products was made for two reasons. The first and practical reason is that drug products consist of a much larger number of different brand names than their pharmaceutical ingredients. A pharmaceutical ingredient, such as Ibuprofen, has only a few different names, while its corresponding drug products are on the market in a much greater variety of names, e.g. Nurofen, Advil, and Nuprin. This makes the identification and normalization in texts much more difficult. Secondly, medical research is more concerned with the basic pharmaceutical ingredients than with commercial drug products from my perspective, which makes the discovery of pharmaceutical ingredients in medical texts the more important task.

Gene Dictionary

The gene dictionary I assembled contains a single entity type called GENE. Under this entity type 122,975 entities, each representing a human gene, are defined. All gene entities have one standard name and a varying number of alternative names.

To obtain the standard and alternative names of all human genes I consulted the GeneCards web directory [252]. It comprises of comprehensive gene information collected from a wide range of data sources like the Database of Genomic Variants (DGV), the HUGO Gene Nomenclature Committee (HGNC) database, the Online Mendelian Inheritance in Man (OMIM) database and PubMed [242, 218, 221, 230]. As a starting point for the dictionary I used the directory's plain text gene list [251]. For all 21,658 protein-coding genes listed on GeneCards as of March 2013, I consulted the genes' GeneCards to obtain all aliases and synonyms per gene. Thus, the dictionary contains a total of 186,771 alternative names, which converts to an average of approx. 8.6 alternative names per protein-coding gene.

I chose GeneCards as data source because its Aliases section is the most comprehensive of all large gene directories [252]. The reason for this is that GeneCards collects and unifies information from numerous gene dictionaries and sources and therefore lists more gene aliases than each of them [253].

Very short gene names like "T", "NA" or "PC" result in more false than right positives due to their ambiguity. A length greater than two, however, makes ambiguity much less probable, which is why I decided to remove gene names shorter than three letters from the dictionary.

Pharmaceutical Ingredient Dictionary

The pharmaceutical ingredient dictionary I assembled contains a single entity type called `PHARMACEUTICAL_INGREDIENT`. As data source for this dictionary I chose Food and Drug Administation's (FDA) Metathesaurus Structured Product Labels (MTHSPL) of the from the Unified Medical Language System (UMLS) [248]. The UMLS is "a set of files and software that brings together many health and biomedical vocabularies and standards to enable interoperability between computer systems" [250]. The MTHSPL is only one of many UMLS sources and lists drug products and pharmaceutical ingredients from the FDA's Structured Product Labeling (SPL). I chose the MTHSPL because it contains official data from the U.S. government's authority for drug administration and because it lists not only drug products but also basic pharmaceutical ingredients.

The UMLS uses a sophisticated classification system to unify multiple names for a single real world entity into a concept, which allowed me to easily built up the dictionary structure of standard form names and alternative names. The dictionary contains 7,099 pharmaceutical ingredients with a total of 5,561 synonyms, which converts to less than one synonym per ingredient.

8.4.2 Customized Extraction Rules

Most clinical trials set age limits for prospective participants. The automatic extraction of these age limits allows the filtering of trials based on a given age. To achieve that I formulated the CGUL rule shown in Listing 8.4.

It defines the entity type `ELIGIBLE_AGE` as follows: The tokens `age` and `eligible` or `allowed` with up to two arbitrary tokens in between and up to three optional tokens followed by:

- **A lower limit**: A number, one arbitrary token (the time unit) and the tokens and and `older`, e.g. 20 years and older,
- **An upper limit**: The tokens up and `to`, a number and one arbitrary token (the time unit), e.g. up to 75 years, or

```
#group ELIGIBLE_AGE (scope='paragraph'):
{
  ((<STEM: age|Age> <>{0,2}
    <STEM: eligible|allowed|Eligible|Allowed>)
    |
    (<STEM: eligible|allowed|Eligible|Allowed> <>{0,2}
    <STEM: age|Age>))
  <>{0,3}
  ((<up> <to> [OD MAX_AGE] <POS:Num> [/OD]
    [OD MAX_AGE_T] <> [/OD])
  |
  ([OD MIN_AGE] <POS:Num> [/OD] [OD MIN_AGE_T] <> [/OD]
    <to> [OD MAX_AGE] <POS:Num> [/OD] [OD MAX_AGE_T] <> [/OD])
  |
  ([OD MIN_AGE] <POS:Num> [/OD] [OD MIN_AGE_T] <> [/OD]
    <and> <older>))
}
```

Listing 8.4: Definition of the entity type `ELIGIBLE_AGE` in CGUL

- **A range limit**: A number, one arbitrary token (first time unit), the token `to`, a number and another arbitrary token (second time unit), e.g. 15 years to 60 years.

The [OD][/OD] tag pairs define and enclose additional entity types within the `#group` directive. Thus, for the string "Ages eligible: 12 weeks to 36 weeks" the entities shown in Table 8.1 will be extracted.

Entity	Type
`Ages eligible: 12 weeks to 36 weeks`	ELIGIBLE_AGE
`12`	MIN_AGE
`weeks`	MIN_AGE_T
`36`	MAX_AGE
`weeks`	MAX_AGE_T

Table 8.1: Identified entities for string "Ages eligible: 12 weeks to 36 weeks".

Most trials also have fixed dates defining the time period in which the trial is active and recruiting participants. The automatic extraction of these dates allows the filtering of trials based on a given day. I formulated the CGUL rules shown in Listing 8.5 to achieve that.

They define the entity type `TRIAL_START` as follows: The tokens `trial` and `start` or `begin` with up to three arbitrary tokens in between, up to two optional tokens, a date consisting of one arbitrary token (the trial start month) and a number (the trial start year). The entity type `TRIAL_END` is defined accordingly. The `[TE DATE][/TE]` tag pairs match a date expression whose structure is defined more precisely between the tags. The `[OD][/OD]` tag pairs define the four en-

```
#group STUDY_START (scope='paragraph'):
{
((<STEM:study|Study> <>{0,3} <STEM:start|begin|Start|Begin>)
|
(<STEM:start|begin|Start|Begin> <>{0,3} <STEM:study|Study>))
<>{0,2}
[TE DATE] [OD STUDY_START_MONTH] <> [/OD]
  [OD STUDY_START_YEAR] <POS:Num> [/OD] [/TE]
}

#group STUDY_END (scope='paragraph'):
{
((<STEM:study|Study> <>{0,3}
  <STEM:end|completion|End|Completion>)
|
(<STEM:end|completion|End|Completion> <>{0,3}
  <STEM:study|Study>))
<>{0,2}
[TE DATE] [OD STUDY_END_MONTH] <> [/OD] [OD STUDY_END_YEAR]
  <POS:Num> [/OD] [/TE]
}
```

Listing 8.5: Definitions of the entity types TRIAL_START and TRIAL_END in CGUL

tity types TRIAL_START_MONTH, TRIAL_START_YEAR, TRIAL_END_MONTH and TRIAL_END_YEAR, which enables the filtering of trials based on their dates.

8.4.3 Post-processing of Text Analysis Results

To start with the indexing of trials their description texts are loaded into the database table TRIALS. Subsequently, a full-text index is created on the text column, which requires text analysis.

As described in Section 8.3 all recognized entities from text analysis are written into a single result table. After indexing has finished this table contains every found gene, pharmaceutical ingredient, trial date and age limit. Each entity is written into a separate row. As a preparation step, I created four database views. Their purpose is to transform the result table into four individual tables where one row represents one trial. The names of the created database views are AGES, DATES, GENES and INGREDIENTS. They contain information of the trial according to their names. This allows the direct access on required information of a certain clinical trial.

Every database view has got the column IDENTIFIER, which is unique for each trial and enables easy joining of the views. The database views transform the results not only structurally but also in terms of content. Extracted numbers are converted from a VARCHAR to an INTEGER data type and names of months are translated into the corresponding number, such as May to 5. This enables the numeric comparison of months and other time values. Additionally, the individual occurrences of genes and pharmaceutical ingredients are grouped and summed up for each trial in order to enable the comparison of trials based on gene or ingredient frequency.

8.4.4 Trial Filtering

The clinical trial search tool now builds upon the `TRIALS` table and the four database views. It expects the following search parameters defining how to identify relevant clinical trials:

- Age of the patient,
- Month and year of planned trial participation,
- (optional) genes the trial should deal with, and
- (optional) pharmaceutical ingredients the trial should deal with.

Genes and pharmaceutical ingredients are optional inputs. I will refer to them as input entities. Matching trials are returned in the following format:

- Unique trial identifier,
- Trial title,
- Trial purpose,
- Month and Year of trial start,
- Month and Year of trial end,
- Minimal participant age in years, months, weeks, days and hours,
- Maximal participant age in years, months, weeks, days and hours,
- Sum of all hits for input entities, and
- Count of how many of the input entities occur in the trial.

The filtering of trials based on the search parameters is achieved with a dynamically synthesized SQL statement. Depending on the existence of genes or pharmaceutical ingredients in the search query the corresponding database views have to be included in the statement or not. Listing 8.6 shows the FROM clause of the statement. Firstly, all trials containing input entity hits are collected from the GENES and INGREDIENTS database views. Then joins with the DATES and AGES views as well as the basic TRIALS table add additional information to the trials, such as trial dates, age limits, the full trial text and the trial title.

The WHERE clause ensures that trials whose dates or age limits do not match the inputs from the search query are removed from the result. Subsequently, the ORDER BY clause sorts the result by:

1. `HIT_COUNT` in descending order,
2. `HIT_SUM` in descending order, and
3. `IDENTIFIER` in ascending order.

This ensures that trials that contain most of the input entities appear at the beginning of the list. If several trials contain the same number of input entities, the sum of all input entity hits decides. Finally, the SELECT clause formats the query output according to what is outlined above.

8.4.5 *User Interface*

The research prototype can be used through the HIG platform's web application and through a tablet application. The web application enables the user to access the trial search tool worldwide on any PC with Internet connection and offers its full functionality. The tablet app, however, targets the mobile use case. It allows users to access the tool even when no PC is within reach.

Web Application

The HIG platform integrates several genome data processing applications, such as an alignment coordinator and a genome browser [237]. The trial search tool is embedded as another application in two places. Firstly, it is displayed contextually next to the list of genetic variants for an uploaded genome sequence as shown in Figure 8.2. At this point clinical trials are found that deal with as many of the displayed varied genes as possible. The search parameters stem from the following sources:

- All genetic variants of the current genome sequence constitute the input genes,
- The age of the patient is looked up in a separate database table that holds patient metadata for the genome sequence files uploaded to HIG, and
- As date of trial participation, the current date is assumed.

As a second possibility, the clinical trial search tool is accessible in a standalone tab of the web application. This tab allows the flexible input of all search parameters and returns all matching trials. The features depicted in Figure 8.3 are accessible from both places. An auto-complete functionality supports the user by displaying genes and pharmaceutical ingredients matching the letters that have already been typed into the search box. The functionality offers completion for all genes and ingredients extracted from the trials. Furthermore, a preview of the trial description appears when the cursor hovers over the trial title. This was implemented in order to enable the user to examine a trial without having to open a new browser window or tab.

Tablet Application

I am convinced that users of the trial search tool, such as physicians or researchers, want to have access to relevant data on mobile devices. A physician, for example, wants to receive matching trials while he is on his way to the next patient. There-

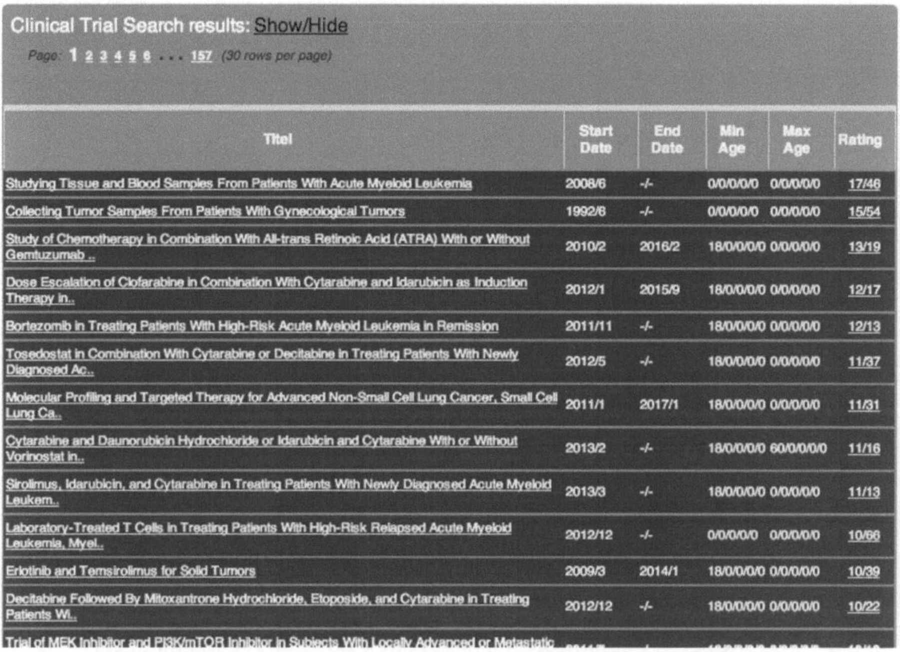

Clinical Trial Search results: Show/Hide

Page: 1 2 3 4 5 6 . . . 157 (30 rows per page)

Titel	Start Date	End Date	Min Age	Max Age	Rating
Studying Tissue and Blood Samples From Patients With Acute Myeloid Leukemia	2008/6	-/-	0/0/0/0/0	0/0/0/0/0	17/46
Collecting Tumor Samples From Patients With Gynecological Tumors	1992/6	-/-	0/0/0/0/0	0/0/0/0/0	15/54
Study of Chemotherapy in Combination With All-trans Retinoic Acid (ATRA) With or Without Gemtuzumab ..	2010/2	2016/2	18/0/0/0/0	0/0/0/0/0	13/19
Dose Escalation of Clofarabine in Combination With Cytarabine and Idarubicin as Induction Therapy in..	2012/1	2015/9	18/0/0/0/0	0/0/0/0/0	12/17
Bortezomib in Treating Patients With High-Risk Acute Myeloid Leukemia in Remission	2011/11	-/-	18/0/0/0/0	0/0/0/0/0	12/13
Tosedostat in Combination With Cytarabine or Decitabine in Treating Patients With Newly Diagnosed Ac..	2012/5	-/-	18/0/0/0/0	0/0/0/0/0	11/37
Molecular Profiling and Targeted Therapy for Advanced Non-Small Cell Lung Cancer, Small Cell Lung Ca..	2011/1	2017/1	18/0/0/0/0	0/0/0/0/0	11/31
Cytarabine and Daunorubicin Hydrochloride or Idarubicin and Cytarabine With or Without Vorinostat in..	2013/2	-/-	18/0/0/0/0	60/0/0/0/0	11/16
Sirolimus, Idarubicin, and Cytarabine in Treating Patients With Newly Diagnosed Acute Myeloid Leukem..	2013/3	-/-	18/0/0/0/0	0/0/0/0/0	11/13
Laboratory-Treated T Cells in Treating Patients With High-Risk Relapsed Acute Myeloid Leukemia, Myel..	2012/12	-/-	0/0/0/0/0	0/0/0/0/0	10/66
Erlotinib and Temsirolimus for Solid Tumors	2009/3	2014/1	18/0/0/0/0	0/0/0/0/0	10/39
Decitabine Followed By Mitoxantrone Hydrochloride, Etoposide, and Cytarabine in Treating Patients Wi..	2012/12	-/-	18/0/0/0/0	0/0/0/0/0	10/22
Trial of MEK Inhibitor and PI3K/mTOR Inhibitor in Subjects With Locally Advanced or Metastatic	[illegible]	[illegible]	[illegible]	[illegible]	[illegible]

Fig. 8.2: The results of the contextual trial search, which are displayed next to the genetic variant list.

fore, the trial search tool was embedded in a tablet application, which offers a subset of the web application's functionality. The app allows the input of genes and pharmaceutical ingredients and displays the resulting trials in a list as shown in Figure 8.4.

8.5 Benchmarks

The research prototype that was introduced in the previous chapter is to be benchmarked based on various measures. The aim is to prove that it can be applied to large real-world data sets and that its performance allows for the growth of input data in future years.

I will focus on the following aspects in my benchmarks.

- How much does the runtime increase when the number of genes is increased?
- Does the performance improve when the text analysis result table is partitioned?
- How much does the runtime grow when the number of trials in the database increases?

The core of the research prototype is the dynamically created SQL statement described in Section 8.4.4 that retrieves trials from the database that matches the given input criteria. Its performance determines the general performance of the

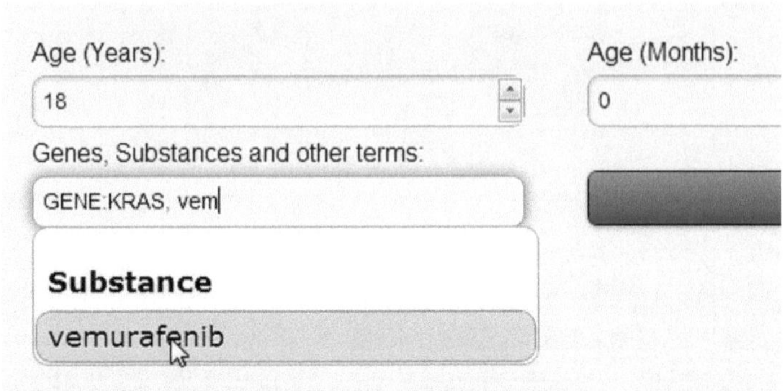

(a) The auto-completion supports while entering valid gene and pharmaceutical ingredient names.

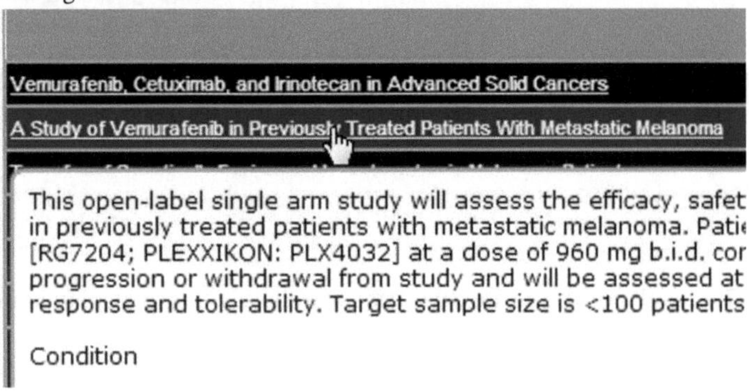

(b) When hovering with the mouse over the trial's title, a preview of its detailed description is provided.

Fig. 8.3: Selected features of the trial search tool.

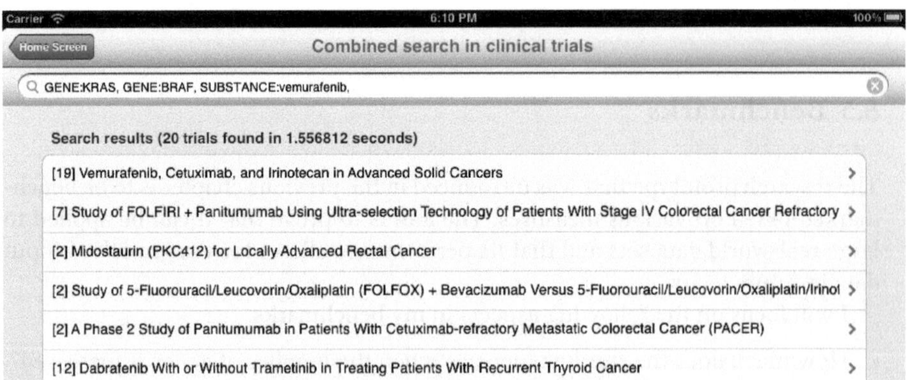

Fig. 8.4: The result list on a tablet.

prototype. Therefore, I confined the benchmarks to measuring the execution time of this SQL statement.

The benchmarks are all done using a custom Python script. Python in version 2.6.4.10 is used and the time measurement is conducted using Python's `timeit` library. The access to the database is established through the Python module `pyhdbcli`, which builds upon the `pyodbc` Python module.

When doing benchmarks one has to consider the measurement accuracy of the measuring instruments. The `timeit` modul, which I use provides a simple way to time small bits of code in Python. It automatically picks an appropriate way of measuring time on each platform and avoids a number of common traps for measuring [233]. As I conduct the benchmarks on a Windows operating system `timeit` uses `time.clock` for measurements. This function is based on the Win32 function `QueryPerformanceCounter()` and its resolution on Windows is better than one microsecond [232].

The test system consists of two identical servers, each with the specification shown in Table 8.2. All tests are run on this system.

Processor	8 x Intel® Xeon® E7-8870 @ 2.40 GHz
Main Memory	64 x 32 GB DDR3 1066 MHz + 64 x 16 GB DDR3 1066 MHz
Operating System	SUSE Linux Enterprise Server 11.2 (Kernel 3.0.13-0.27-default)

Table 8.2: Configuration of benchmark system.

8.5.1 Genes

The first benchmark aims to explore the increase in execution time dependent on the number of genes given to the query. It can give an insight into how the execution time grows when the genetic variant list is large and the trial search is called on these genes. For the benchmark, I imported all 30,408 recruiting trials as of May 2013 from clinicaltrials.gov into the database because these are the trials most relevant to patients who wish to participate in one. To get a comprehensive impression of the prototype's performance I examine a range of 1 to 4480 genes. 4480 is the total number of genes recognized in all recruiting trials from clinicaltrials.gov and is therefore assumed as maximum value. As gap between two measurement settings I set 100 genes with 1 instead of 0 as beginning point.

For each of these measurement settings, 20 values are measured by executing the query 20 times. Data has been cleansed by using measurements only within the 90 % confidence interval. Then the standard deviation over the remaining values is calculated and compared with the target value of five percent. If this target is missed, additional measurements are taken in groups of ten until the standard deviation is equal or less than five percent.

8.5.2 *Partitioning the Text Analysis Result Table*

The second benchmark aims to determine whether the execution time of the query benefits from a partitioned text analysis result table. It can give an insight into how the performance of the trial search can be improved.

The partitioning schema I use is hash-range partitioning. On the first level I create two hash partitions on the `IDENTIFIER` column of the result table and distributed them on the two servers. On the second level, I create eleven range partitions on the `TA_TYPE` column containing the following values:

- `MIN_AGE`,
- `MIN_AGE_T`,
- `MAX_AGE`,
- `MAX_AGE_T`,
- `TRIAL_START_MONTH`,
- `TRIAL_START_YEAR`,
- `TRIAL_END_MONTH`,
- `TRIAL_END_YEAR`,
- `GENE`,
- `PHARMACEUTICAL_INGREDIENT`, and
- all other values.

This assigns the extracted entities based on their type to the partitions. While the large final partition holds all those entities that are not relevant to the trial search, the other 10 partitions are relatively small and should therefore be quick to read.

For the benchmark, I follow the same measurement procedure with 45 different gene numbers as in the first benchmark in order to allow a comparison with the values of the first benchmark.

8.5.3 *Indexed Trials*

The third benchmark aims to explore the increase in execution time dependent on the number of clinical trials indexed in the database. It provides insight into how in the future execution time increases as more and more recruiting trials are registered in online directories, such as clinicaltrials.gov. In this benchmark, I examine a range of 10,000 to 60,000 clinical trials and set a gap of 10,000 trials between two measurement settings. Benchmarking data was obtained from clinicaltrials.gov and is randomly chosen from all available trials. I configure the query with an input gene number of 4,400. This number is the highest value from the test settings of the benchmarks 1 and 2 and ensures that the difference between the taken measurement values is significant.

For each of the six measurement settings 20 values are measured and a confidence interval of 90 % is chosen. The target value for the relative standard deviation remains 5 % and additional measurements are taken in groups of ten until the standard deviation falls below this level.

8.6 Discussion

In the following, I discuss the results obtained in each benchmark. Due to the relatively small standard deviations, only 20 values were measured for each setting in all three benchmarks.

Number of Given Genes

The trial search tool was designed for the application on patient-specific genome data. I am convinced that by using the genomic variants of a patient as input to the tool the results will be more relevant and apposite than when using manually selected genes as input. The tool will be aware of all altered genes in the patient's genome and can list these trials that deal with most of these genes.

The genetic variant list, however, can contain very many genes. Therefore, this benchmark has aimed at exploring the increase in execution time dependent on the number of input genes. Figure 8.5 shows a box plot to visualize the distribution of the measurement values. The figure not only shows values from the confidence interval but all measured values. The boxes represent the differences between the first and third quartiles with horizontal lines at the median values. Whiskers extend from the boxes to the most distant points, whose y-values are within 1.5 times the interquartile ranges. Points outside these limits are drawn individually.

The box plot reflects the low standard deviation of maximally three percent. The interquartile range is constantly and significantly smaller than 0.5 seconds and very few outliers exist. This indicates that the benchmark gives a reliable impression of the realistic performance of the system.

A clear upward trend can be seen when the number of genes increases. More genes contained in the searched query lead to a larger subset of trials dealing with at least one of them and consequently more trials are fetched from the GENES database view. Subsequently, these trials are joined with the DATES and AGES views and the TRIALS table, they are filtered by trial dates and age limits, and finally ordered. All of these operations need to be performed on more data when the number of input genes increases, which is reflected in the plot.

Additionally, it is visible that the execution time is not growing linearly but less than that. Figure 8.5 shows a square root function laid over the box plot. The progression of the execution time follows the square root function very tightly, which prompts the conclusion that the execution time of the query grows similar to a square root function.

This is very beneficial to the trial search tool, as the execution time will increase more slowly than the input size and larger variant lists can be handled without much impact on performance.

The plot shows an execution time ranging from 2.0 s to 4.4 s. I value these results as promising for such a compute-intense task, which can be performed now within an interactive response time. Nevertheless, I am convinced that the performance of the query can be optimized in order to reduce the response time of the trial search tool further, e.g. by applying appropriate indices.

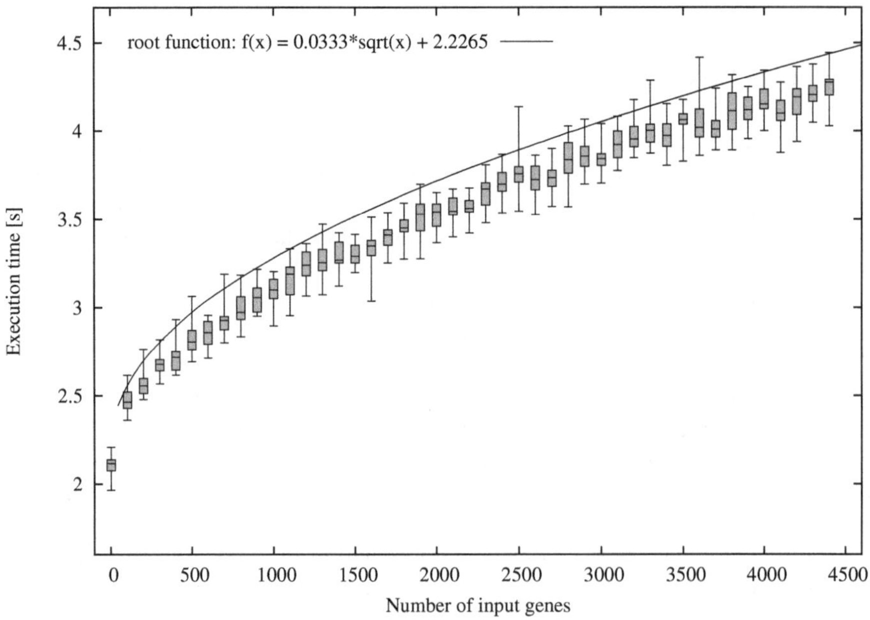

Fig. 8.5: Execution time increases in an logarithmic manner when input gene number increases.

Partitioning on the Text Analysis Result Table

Figure 8.6 compares the measurements taken in the first benchmark with the results from this benchmark. The points represent the mean of the measurement values in the confidence interval. The solid line connects the execution times measured without partitioning. The dashed line represents the execution times measured with partitioning enabled.

It can be clearly seen that the solid and the dashed curve follow a similar progression. Furthermore, the performance gained by enabling the partitioning is ap-

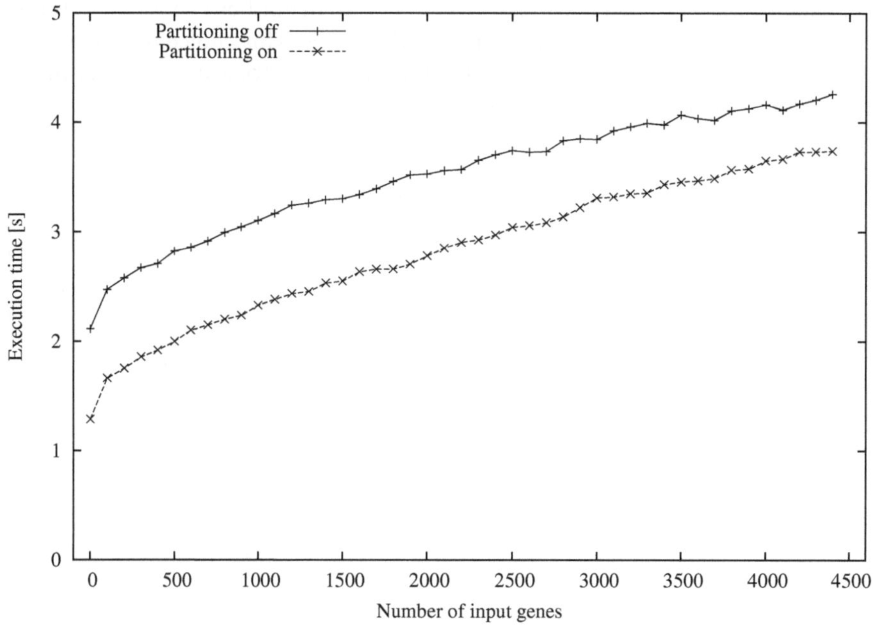

Fig. 8.6: Partitioning reduces execution times significantly.

parent, too. I consider this difference as significant but also improvable. The two-level partitioning splits the text analysis result table into 22 partitions on 2 servers. Nevertheless, some partitions are still very large and can only be accessed by one search thread at a time. Particularly the GENES and PHARMACEUTICAL_INGREDIENT partitions should be partitioned further. Although third-level partitioning is not possible with the IMDB I used, I am convinced that it would increase the performance further. It would enable the parallel access of multiple genes and ingredients, which I consider as a good strategy to accelerate the query.

Number of Indexed Trials

It can be expected that in future years the number of clinical trials registered in online databases will rise. As of June 20, 2013, clinicaltrials.gov lists already more than 147,000 trials and adds 400 new records each week [254]. Consequently, a tool aiming to search all recruiting trials must be able to handle a growing number of indexed trials.

This benchmark aimed to explore the increase in the execution time of the trial search query dependent on the number of indexed trials. Figure 8.7 shows a box plot to visualize the distribution of the measurement values. The figure shows all measured values and the fitted function $f(x) = 0.0000028 * x + 1$. Although the benchmark was designed to compare different trial quantities I chose the total size of the trials on disk in kilobytes as unit for the x-axis. This choice reflects the fact that trials have various sizes and that the total size of all trials has a greater impact on the query performance than their number.

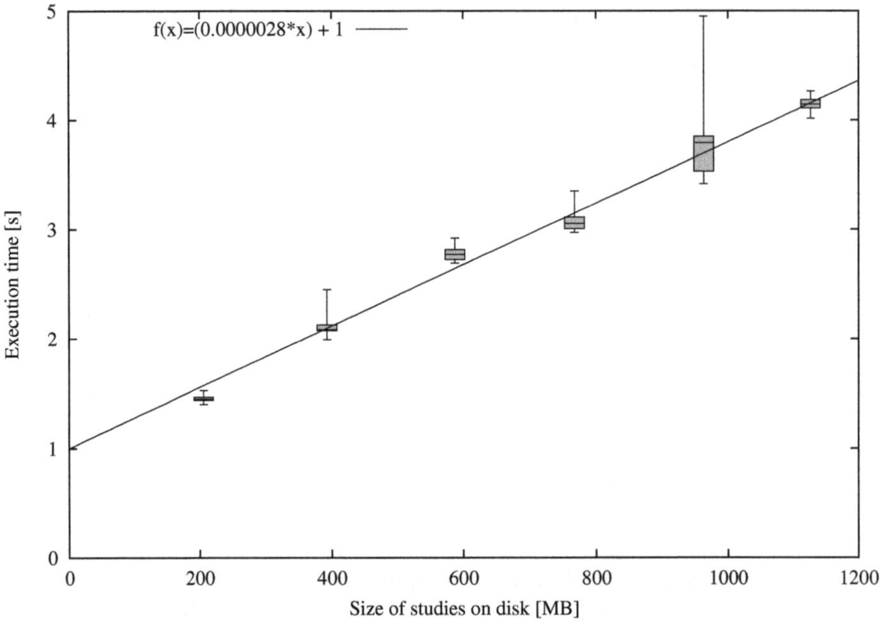

Fig. 8.7: Execution time increases approximately proportionally to the total size of trials on disk.

The box plot reflects this benchmark's low standard deviation of maximally 4.42 %. The interquartile range is constantly smaller than 0.5 seconds and very few outliers exist. This indicates that the benchmark shows the realistic performance of the system. It is clearly visible that the execution time grows with greater trial size and that this growth correlates with the overlaid linear function. This can be explained with the linear dependence between the amount of indexed text and the amount of text analysis results to be examined during the query.

In the first benchmark, 30,408 trials with a total size of 711,608 KB were indexed by the database and the mean time in the confidence interval for 4,400 genes was 4.260 s. However, the third benchmark measured a significantly lower mean time of 3.061 s for 4,400 genes and a total trial size of about 767,628 KB in 40,000 trials. Although these values seem to contradict each other, it has to be considered that they were measured in different and independent benchmarks. The third benchmark operated on randomly chosen trials, whereas the first benchmark operated on the specific set of all recruiting trials. This difference in the setup could have involved different execution plans for the query, which would explain the time difference.

8.7 Conclusion and Outlook

My work addressed the problems occurring when structured and unstructured medical data is examined separately. I am convinced that structured patient data enables a much more targeted search in unstructured natural language texts and that a combined search of both data types can benefit physicians and patients.

In this work, I showed how IMDB technology can be utilized for such a search task. I outlined the text analysis features IMDBs offer and explained how they can be configured for the biomedical domain. Furthermore, I shared a detailed insight into my research prototype, which filters clinical trials based on patient-specific genomic data and metadata. I described customized dictionaries and extraction rules that enable IMDBs to extract trial information as well as names of genes and pharmaceutical ingredients from unstructured natural language texts. As a result, my research prototype is able to retrieve and analyze information from trial descriptions stored in an IMDB. I depicted the architecture of the research prototype and presented its user interface that enables access both through a web service and a tablet application.

Ultimately, I executed three individual benchmark setups to evaluate the prototype's performance when using different sets of medical data. The benchmarks revealed that the execution time increases logarithmically with the number of genes in the input and that partitioning can optimize the prototype's performance. Additionally, the benchmarks suggested a linear correlation between execution time and the amount of indexed trial data. The results indicate that the research prototype benefits from the IMDB's ability to handle large amounts of data and that its performance allows for the growth of medical input data in future years.

I expect future work to extract additional details from the trial descriptions, so that the result set is extended.

In general, I expect that other scientific projects will continue with the combination of structured and unstructured medical data. IMDBs can be used to search trial descriptions and other types of natural language texts, such as diagnosis documents. Consequently, applications can be developed that examine sources, such as biomedical publications or medical records, and find the relevant documents for a patient. The scientific community has explored the individual search in either structured or unstructured medical data for several years. However, I identified their combination as a new research topic and I am convinced that its analysis will provide additional insight for medical experts.

8.8 Appendix

This appendix contains the SQL code for filtering from my prototype, which is shown in Listing 8.6.

```sql
FROM
(
  (
  -- select study identifier, sum of all input entity hits and
     count of how many of the input entities occur in the study
  SELECT
    IDENTIFIER,
    SUM(NUM) AS HIT_SUM,
    COUNT(IDENTIFIER) AS HIT_COUNT
  FROM
  (
    (
    -- retrieve hits of input genes
    SELECT
      IDENTIFIER,
      GENE AS NAME,
      NUM
    FROM
      -- contains gene hits in the format (identifier, gene, hit
         number) with one row for each identifier-gene pair
      SEARCH.GENES
    WHERE
      -- selection only of hits for input genes
      GENE IN (<genelist>)
    )
    -- union with hits of input ingredients
    UNION
    (
    SELECT
      IDENTIFIER,
      INGREDIENT AS NAME,
      NUM
    FROM
      -- contains ingredient hits in the format (identifier,
         ingredient, hit number) with one row for each
         identifier-ingredient pair
      SEARCH.INGREDIENTS
    WHERE
      -- selection only of hits for input ingredients
      INGREDIENT IN (<ingredientlist>)
    )
  )
  -- group all input entity hits for a study; this enables
     adding up and counting of hits
  GROUP BY IDENTIFIER
  ) AS GROUPED_TERM_TABLE
  -- add study dates to the studies
  LEFT JOIN
  SEARCH.DATES
  ON GROUPED_TERM_TABLE.IDENTIFIER = SEARCH.DATES.IDENTIFIER
  -- add age limits to the studies
  LEFT JOIN
  SEARCH.AGES
  ON GROUPED_TERM_TABLE.IDENTIFIER = SEARCH.AGES.IDENTIFIER
  -- add study text and title to the study
  INNER JOIN
  SEARCH.STUDIES
  ON GROUPED_TERM_TABLE.IDENTIFIER = SEARCH.STUDIES.IDENTIFIER
)
```

Listing 8.6: SQL FROM clause of the filtering statement

8.9 References

[214] Boese JH et al. (2012) Data Management with SAP's In-memory Computing Engine. In: Proceedings of the 15th International Conference on Extending Database Technology

[215] Chang JT, Schütze H, Altman RB (2004) GAPSCORE: Finding Gene and Protein Names One Word at a Time. Bioinformatics Journal 20(2):216–225

[216] Chiang JH, Yu HC (2003) MeKE: Discovering the Functions of Gene Products from Biomedical Literature via Sentence Alignment. Bioinformatics Journal 19(11):1417–1422

[217] Cios KJ, William Moore G (2002) Uniqueness of medical data mining. Artificial intelligence in medicine 26(1):1–24

[218] Committee HGN (2013) HUGO Gene Nomenclature Committee. http://www.genenames.org/. Accessed Sep 23, 2013

[219] DeWitt DJ et al. (1984) Implementation Techniques for Main Memory Database Systems. In: Proceedings of the International Conference Management of Data, ACM, pp 1–8

[220] Garcia-Molina H, Salem K (1992) Main Memory Database Systems: An Overview. IEEE Transactions on Knowledge and Data Engineering 4(6):509–516

[221] Hamosh A et al. (2005) Online Mendelian Inheritance in Man (OMIM), a Knowledgebase of Human Genes and Genetic Disorders. Nucleic Acids Research 33:D514 – D517

[222] Hunt DL et al. (1998) Effects of Computer-based Clinical Decision Support Systems on Physician Performance and Patient Outcomes. Journal of the American Medical Association 280(15):1339–1346

[223] Ibrahim GM, Chung C, Bernstein M (2011) Competing for Patients: An Ethical Framework for Recruiting Patients with Brain Tumors into Clinical Trials. Journal of Neuro-Oncology 104(3):623–627

[224] Kanehisa M, Goto S (2000) KEGG: Kyoto Encyclopedia of Genes and Genomes. Nucleic Acids Research 28(1):27–30

[225] Knöpfel A, Gröne B, Tabeling P (2005) Fundamental Modeling Concepts. Wiley, West Sussex UK

[226] Krallinger M, Valencia A (2005) Text-mining and Information-retrieval Services for Molecular Biology. Genome Biology 6(7):224

[227] Krallinger M et al. (2008) Evaluation of Text-mining Systems for Biology: Overview of the Second BioCreative Community Challenge. Genome Biology 9 supplement 2:S1

[228] Krallinger M et al. (2008) Linking Genes to Literature: Text Mining, Information Extraction, and Retrieval Applications for Biology. Genome Biology 9, supplement 2:S8

[229] Nadeau D, Sekine S (2007) A Survey of Named Entity Recognition and Classification. Lingvisticae Investigationes 30(1):3–26

[230] National Center for Biotechnology Information, U.S. National Library of Medicine (2013) Pubmed. http://www.ncbi.nlm.nih.gov/pubmed. Accessed Sep 23, 2013
[231] Plattner H (2013) A Course in In-Memory Data Management: The Inner Mechanics of In-Memory Databases. Springer
[232] Python Software Foundation (2013) 15.3. Time - Time Access and Conversions - Python v2.7.5 documentation. http://docs.python.org/2/library/time.html. Accessed Sep 23, 2013
[233] Python Software Foundation (2013) 26.6. Timeit - Measure Execution Time of Small Code Snippets. http://docs.python.org/2/library/timeit.html. Accessed Sep 23, 2013
[234] SAP AG (2013) SAP HANA Developer Guide. http://help.sap.com/hana/SAP_HANA_Developer_Guide_en.pdf. Accessed Sep 23, 2013
[235] SAP AG (2013) Text Data Processing Extraction Customization Guide. http://help.sap.com/businessobject/product_guides/sboDS42/en/ds_42_tdp_ext_cust_en.pdf. Accessed Sep 23, 2013
[236] SAP AG (2013) Text Data Processing Language Reference Guide. http://help.sap.com/businessobject/product_guides/boexir4/en/sbo401_ds_tdp_lang_ref_en.pdf. Accessed Sep 23, 2013
[237] Schapranow MP, Plattner H, Meinel C (2013) Applied In-Memory Technology for High-Throughput Genome Data Processing and Real-time Analysis. In: Proceedings of the XXI Winter Course of the Centro Avanzado Tecnológico de Análisis de Imagen, pp 35–42
[238] Schapranow MP et al. (2013) Mobile Real-time Analysis of Patient Data for Advanced Decision Support in Personalized Medicine. In: Proceedings of the 5th International Conference on eHealth, Telemedicine, and Social Medicine
[239] Settles B (2005) ABNER: An Open Source Tool for Automatically Tagging Genes, Proteins and other Entity Names in Text. Bioinformatics Journal 21(14):3191–3192
[240] Sittig DF et al. (2008) Grand challenges in clinical decision support v10. Journal of biomedical informatics 41(2):387
[241] Tanabe L, Wilbur WJ (2002) Tagging Gene and Protein Names in Full Text Articles. In: Proceedings of the Workshop on Natural Language Processing in the Biomedical Domain, vol 3, pp 9–13
[242] The Centre for Applied Genomics (2013) Database of Genomic Variants. http://dgvbeta.tcag.ca/dgv/app/downloads. Accessed Sep 23, 2013
[243] UniProt Consortium (2013) Universal Protein Resource (UniProt). http://www.uniprot.org/. Accessed Sep 23, 2013
[244] U.S. Food and Drug Administration (2012) The FDA's Drug Review Process: Ensuring Drugs Are Safe and Effective. http://www.fda.gov/drugs/resourcesforyou/consumers/ucm143534.htm. Accessed Sep 23, 2013
[245] U.S. National Institutes of Health (2013) ClinicalTrials.gov. http://www.clinicaltrials.gov/. Accessed Sep 23, 2013

[246] U.S. National Institutes of Health (2013) How to Use Advanced Search - ClinicalTrials.gov. http://clinicaltrials.gov/ct2/help/how-find/advanced. Accessed Sep 23, 2013
[247] U.S. National Institutes of Health (2013) Learn About Clinical Studies - ClinicalTrials.gov. http://clinicaltrials.gov/ct2/about-studies/learn. Accessed Sep 23, 2013
[248] U.S. National Library of Medicine (2013) 2012AB FDA Structured Product Labels Source Information. http://www.nlm.nih.gov/research/umls/sourcereleasedocs/current/MTHSPL/. Accessed Sep 23, 2013
[249] U.S. National Library of Medicine (2013) Citations Added to MEDLINE by Fiscal Year. http://www.nlm.nih.gov/bsd/stats/cit_added.html. Accessed Sep 23, 2013
[250] U.S. National Library of Medicine (2013) Unified Medical Language System (UMLS). http://www.nlm.nih.gov/research/umls/. Accessed Sep 23, 2013
[251] Weizmann Institute of Science (2013) All GeneCards genes. http://genecards.org/cgi-bin/cardlisttxt.pl. Accessed Sep 23, 2013
[252] Weizmann Institute of Science (2013) GeneCards - Human Genes | Gene Database | Gene Search. http://genecards.org/. Accessed Sep 23, 2013
[253] Weizmann Institute of Science (2013) Information Page for GeneCards Sections. http://genecards.org/info.shtml. Accessed Sep 23, 2013
[254] Zarin D et al. (2013) ClinicalTrials.gov and Related Projects: Improving Access to Information about Clinical Trials; A Report to the Board of Scientific Counselors. Technical Report TR-2013-001, Lister Hill National Center for Biomedical Communications, U.S. National Library of Medicine

Chapter 9
Real-time Collaboration in the Course of Personalized Medicine

Hasso Plattner and Matthieu-P. Schapranow

In the following, we outline innovative applications that build real-time analysis of patient data in course of personalized medicine. They describe a completely new way of exploring treatment-relevant information. We focus on the involved business entities and their actions. Table 9.1 classifies processes, involved entities, and required application perspectives.

Business Process	Group	Category
Real-time Combination of Oncology Data	C, R	E
Building Research Hypotheses	R	N
Pharmaceutical Feedback Loop	R	N
Federal Bureau of Statistics	C, R	N
Health Insurance Companies	B	N
Tumor Board of the Future	C, R	E

Table 9.1: Classification of innovative applications and affected user groups (C = Clinicians, R = Researchers, B = Business managers) and category (N = New, E = Extended).

9.1 Real-time Combination of Oncology Data

In the following, we share details about our research prototype Oncolyzer and its application perspectives as summarized in Table 9.2. The system architecture and its integration into the Hospital Information System (HIS) are modeled in Figure 9.1. The key component of the architecture is the IMDB HANA that enables real-time statistical analysis of patient cohort data, medical actions, and data from further clinical systems. The mobile devices are connected via a device manage-

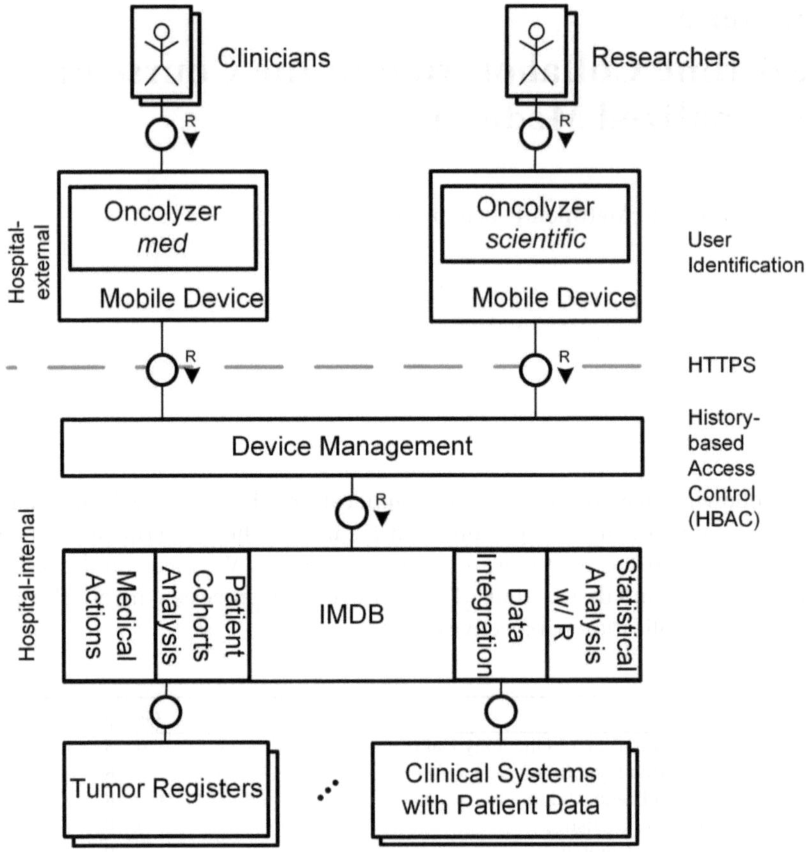

Fig. 9.1: Oncolyzer system architecture: Data from clinical systems is combined and analyzed directly in the IMDB. The results are accessible by the mobile application.

ment layer to the database management system, which combines data from clinical data sources, such as tumor registers or HIS.

Perspective	Group	Sect.
Holistic Patient View	P, R	9.1.1
Search in Structured and Unstructured Data	P	9.1.2
Real-time Analysis of Patient Cohorts	R	9.1.3

Table 9.2: Application perspectives and corresponding user groups (P = Physicians, R = Researchers).

9.1.1 Holistic Patient View

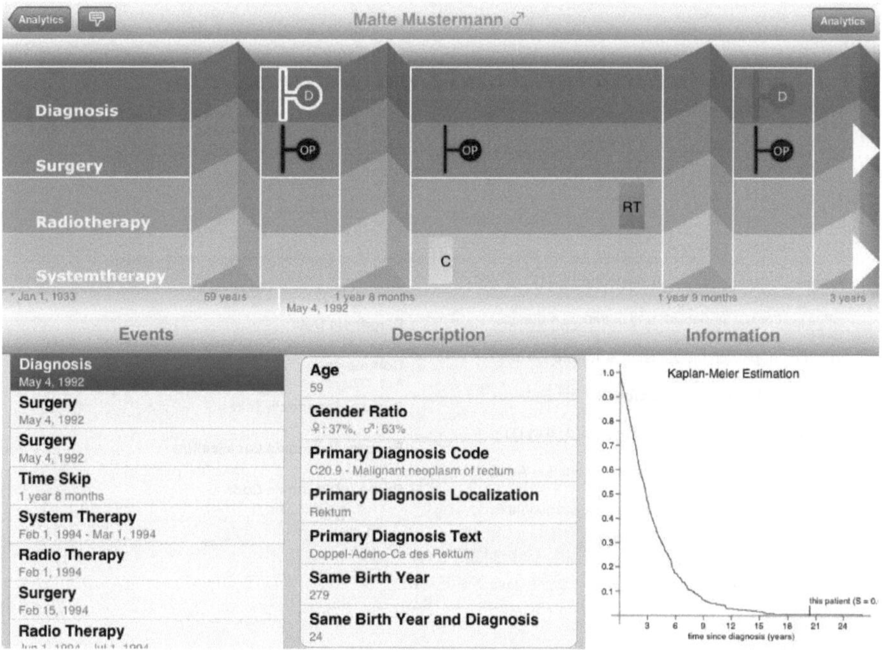

Fig. 9.2: The holistic patient view consists top-down of name and gender details, interactive treatment history combining all relevant treatment information on a single screen as well as a tabular view. The latter contains from left to right details for treatment events, analytical results, and a graphical evaluation of patients with the same primary diagnosis using the Kaplan Meier analysis.

The holistic patient view as depicted in Figure 9.2 combines patient specific data from various clinical databases. For example, the complete treatment history consisting of diagnoses, surgeries as well as radio and system therapies are combined and visualized as a graphical timeline. The IMDB technology performs real-time analysis to identify similarities in the data of the selected patient and data of hundreds or thousands or even hundreds of thousands similar patients. For example, results for patients with the same year of birth, with the same primary diagnosis, or gender ratio are transparently calculated for any selected patient. This environmentally derived information can significantly help to identify sources of diseases, e.g. to identify the connection between TP53 gene mutations and urinary bladder cancer after the nuclear plant catastrophe in Chernobyl in 1986 [263].

The incorporated IMDB is hosted by the central IT department and runs on multiple high-end servers. In contrast to data warehouse systems, which store pre-aggregated totals to improve long-running queries, IMDB technology performs all calculations on the fly. In other words, once a new patient record has been added to the database, it can be accessed immediately via the Oncolyzer. The new record directly influences calculations and analysis results, e.g. Kaplan Meier analysis.

9.1.2 Search in Structured and Unstructured Data

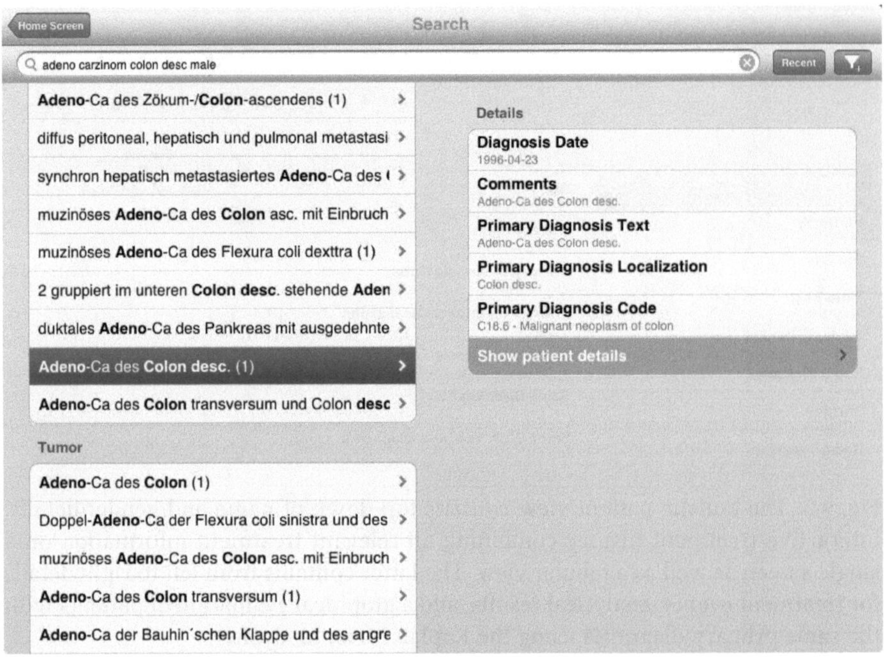

Fig. 9.3: Result set of the search query for "adeno carzinom colon desc male". The search term "carzinoma" is identified by the fuzzy search as equivalent for "carcinoma" and is also substituted by its abbreviation "ca". The result set is also filtered using the value "male" obtained from a search in the structured attribute "gender".

We are convinced that diagnostic reports contain valuable information, although they consist of less standardized free-text documents. As a result, it is hard to access relevant details, e.g. to determine that the subscribed drug was Cetuximab. The Oncolyzer mobile application is equipped with a combined search for structured and unstructured data, which is supported by the underlying IMDB technology. The application perspective consists of a search box following the Google-like UI. Patient documents from different systems, e.g. diagnosis reports, preliminary diseases, death causes, etc., are searched for by the terms entered as depicted in Figure 9.3.

In addition to an exact match search, fuzzy search also detects similar search terms, e.g. in case of typos [257]. Our research showed that fuzzy search is valuable in clinical contexts as medical terms can be documented in Latin, English or German or their abbreviations. Fuzzy search accepts a specific degree of fuzziness in the search term, i.e. the result set contains more relevant search results than an exact search for terms.

We further integrated a synonym search, which returns documents that contain synonyms instead of the searched terms. For instance, the abbreviation "ca." is synonymously used for "circa", but it refers to "carcinoma" in clinical documents. A clinical classifications was added to the synonym table, e.g. the ICD and the corresponding tumor location [266]. Thus, a search for the tumor location "rectum" also returns results containing the ICD code "C20".

Furthermore, stop words for structured search are derived from all values stored as structured attributes. A subsequent search for each stop word is performed in the database table containing the structured attribute. For example, a search for "male adeno ca" is expanded to a combined search for patients having the value "m(ale)" in structured attribute "gender" and associated documents with indications for an adeno-carcinoma.

The combination of search in structured and unstructured data leads to more accurate result sets than traditional search tools, e.g. exact match search.

9.1.3 *Real-time Analysis of Patient Cohorts*

Researchers are able to perform individual real-time analysis on patient data on their mobile devices. The analytical view of our Oncolyzer provides the graphical UI for visualization of results of complex analytical queries that are executed by the IMDB system. Thus, researchers are able to identify relevant correlations with external factors, e.g. to prevent recurrences or to form patient cohorts. The quantity of patients that fulfill specific preconditions can be determined by using individual criteria for filtering, such as the patient's age, the gender, kind of tumor, or its localization as depicted in Figure 9.4. The results are interactively visualized using individual chart types, such as bar, pie or line charts. Clicking on a fragment of the chart shows the list of actual patients. All patients are linked to their holistic patient view, which connects analysis of cohorts and specific patients.

9.2 Building Research Hypotheses

Clinical researchers require flexible ways to test hypotheses and correlations between certain aspects of patient data. BW or BI systems enable the analysis of patient cohorts in a fast way, but involve complex administrative operations for data preparation before accessing results. After patient data is Extracted, Trans-

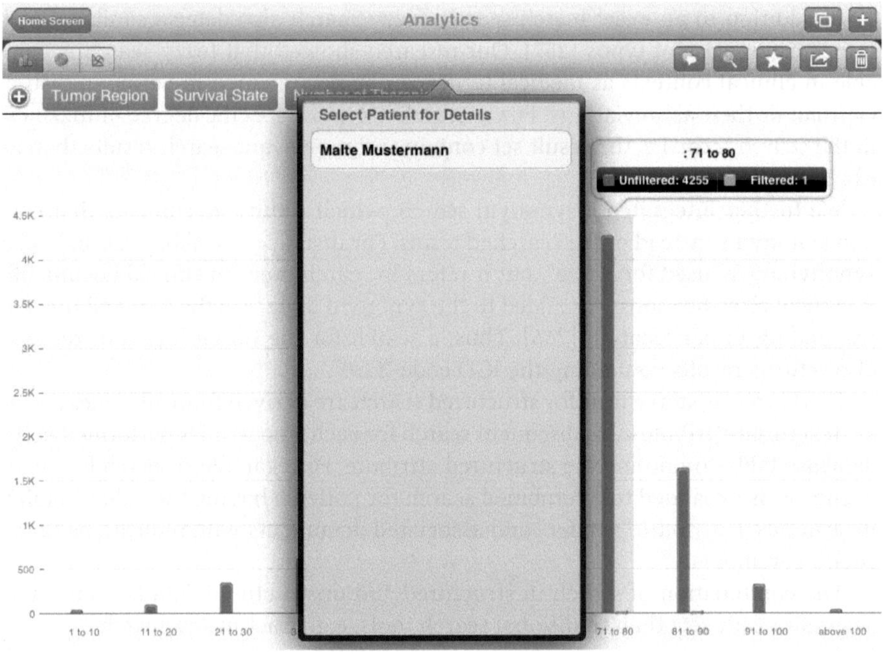

Fig. 9.4: Real-time analysis of patient cohorts: Filters for tumor region, survival state, and number of therapies were applied and a patient in the age group "70 to 80" with the alias "Malte Mustermann" was identified.

formed, and Loaded (ETL) into the Business Intelligence (BI) system [265], they are optimized for predefined queries, i.e. expected correlations need to be included in the reports. We introduce a completely new way of analyzing patient data while bypassing ETL processes. We keep all relevant patient data in the main memory in a columnar format without the need to load and prepare data, e.g. from disks, before starting their analysis. As a result, data is accessed at main memory latency time, i.e. 100 ns, instead of about 10 ms required for hard disk access [264].

9.3 Pharmaceutical Feedback Loop

New pharmaceutical products require to pass clinical trials before being approved for regular use [259]. Pharmaceutical manufacturers require patients with very specific indications to participate in clinical trials. Pharmaceutical researchers can analyze patient cohorts in a similar way as clinical researchers use the analytical view. Their access is restricted to anonymized cohort data only, i.e. they are prohibited to access personal details of the holistic view. If a certain number of patients with similar indication is required, but not present, they can use bookmarks to get notifications once relevant patients are present. Instead of performing these analyses multiple times, a push notification automatically indicates new relevant patients. If the patient should be recruited, the pharmaceutical researcher contacts the hospital and her/his physician informs the patient. Only if the patient agrees, the contact with the pharmaceutical researcher can be established.

9.4 Federal Bureau of Statistics

Statistical analyses of cancer diseases are regularly conducted to supervise influencing factors, such as compliance of treatments with clinical guidelines, rate of new or recurrent cases, and an analysis of regional accumulations [260]. The primary step is to obtain relevant data from nationwide hospitals. In Germany, there are clinical tumor registers that contain well-documented data about all recent cancer cases. A widely used documentation system is the Gießener Tumor Documentation System (GTDS) that provides interactive tools for documentation, verification, and export of analysis data [255]. In future, we expect the following trends to come up:

- The level of detail and therefore the size of the documented data will increase and
- Consolidated nationwide tumor registers will offer new sources of information, e.g. for research [256].

Combining data from nation-wide tumor registers improves the quantity of available data for evidence-based treatment decisions. We implemented a Kaplan Meier analysis that visualizes fractions of patient cohorts and their survival time after first diagnosis [261, Chapter 9]. It is based on patient cohorts with the same ICD of a single hospital. The outlined business process describes a completely new way to combine data from decentralized tumor registers. In future, the basis for these analyses will include data from national and worldwide tumor registers. Thus, the quality of data used for personalized treatment is improved by combining national and international data sources.

Our contributions prove that IMDB technology can be used to bridge individual data formats without the need of data transformation. Therefore, we make use of database views, i.e. a defined transformation rule that is processed, when the underlying data item is accessed [264]. As a result, views are a transparent way to expose a homogenous data format while original data remains unchanged. On the one hand, views require a portion of designated processing time of the CPU. On the other hand, the use of views reduces the time for integration of new data, i.e. required transformations are eliminated. This is beneficial, if you access only a small portion of data and instant transformations are performed (e.g. 100 relevant patients) while traditional ETL processes need to transform the full data (e.g. of 100k patients).

Exposing analytical access to patient data supports employees of the Federal Bureau of Statistics to create national cancer reports. IMDB technology can reduce the integration effort for combining patient data from clinical tumor registers of different vendors. Furthermore, it enables interactive data exploration to trace any kind of statistical anomalies and to determine their natures. Nowadays, tracing anomalies in the reported data is a time-consuming task since experts of the statistics bureau, of the clinic and the physician who have treated a specific patient, need to be involved. As a result, the release cycles of cancer reports can be reduced, for comparison, the German cancer report 2007/2008 was released in 2012 [260].

9.5 Health Insurance Companies

Health insurance companies are interested in optimizing the overall time of hospital stays. However, the duration of a specific stay depends on various factors, such as age of the patient, stage of cancer, etc. Periodically, health insurance companies negotiate fixed rates for treatments of specific disease types with hospitals [258]. The rates are aligned with the average national costs for treatment of the specific disease. However, hospitals specialized in the treatment of certain disease types argue that they receive more complex cases compared to small, regional hospitals. As a result, hospitals are also interested in providing in-house analysis of certain disease types to negotiate more adequate rates. Latest analyses of treated patient cohorts suffering from a specific disease type form the basis for the negotiating process. In addition, it enables a more transparent and holistic view on the treatment process, e.g. to match specific thresholds for the duration of a hospital stay.

9.6 Tumor Board of the Future

In the course of cancer treatments, dedicated experts discuss each individual case. They develop a specific treatment plan, discuss alternatives, and regularly evaluate the performance of the chosen therapy, which requires all relevant data of the discussed patient [262]. All decisions of the tumor board need to be documented and communicated, e.g. new treatment decisions to anybody involved in the treatment process. Nowadays, tumor boards require excessive preparation and paper work to discuss selected patient cases. Participants are limited to experts form a selected discipline or department.

A future way of collaboration within a tumor board is depicted in Figure 9.5. International experts are able to discuss selected patient cases in a virtual conference room via the Internet. Latest data and any kind of ad-hoc analysis can be performed during the discussion to answer questions raised during the expert's debate. As a result, new or missing information can be added instantly during the tumor board session without any latency. Experts from various disciplines can be

Fig. 9.5: Tumor board of the future: international real-time collaboration.

invited to participate in the discussion, e.g. via video conferencing, telephone call or personally. Even the patient can be invited to the tumor board and response to questions from the experts directly. The patient does not need to leave the bed; he is included using the tablet device provided by a clinician for this case.

Having latest data at hand, is a requirement to derive most efficient therapy decisions. As a result, specific analysis tools will assist experts in the tumor documentation.

9.7 References

[255] Altmann U, Katz F, Dudeck J (2012) Das Gießener Tumordokumentationssystem. Spiegel der Forschung
[256] Deutscher Bundesrat (2012) Entwurf eines Gesetzes zur Weiterentwicklung der Krebsfrüherkennung und zur Qualitätssicherung durch klinische Krebsregister (Krebsfrüherkennungs- und -registergesetz - KFRG). http://dipbt.bundestag.de/dip21/brd/2012/0511-12.pdf. Accessed Sep 23, 2013
[257] Färber F et al. (2012) SAP HANA Database: Data Management for Modern Business Applications. Sigmod Record 40(4):45–51
[258] Frakt A (2010) The Future of Health Care Costs: Hospital-insurer Balance of Power. Expert Voices
[259] Ibrahim GM, Chung C, Bernstein M (2011) Competing for Patients: An Ethical Framework for Recruiting Patients with Brain Tumors into Clinical Trials. Journal of Neuro-Oncology 104(3):623–627
[260] Kaatsch P et al. (2012) Krebs in Deutschland 2007/2008. http://www.krebsdaten.de/Krebs/DE/Content/Publikationen/Krebs_in_Deutschland/kid_2012/krebs_in_deutschland_2012.pdf?__blob=publicationFile. Accessed Sep 23, 2013
[261] Lang TA, Secic M (2006) How to Report Statistics in Medicine: Annotated Guidelines for Authors, Editors, and Reviewers. Medical Writing and Communication, ACP Press
[262] Lehmann K et al. (2008) Interdisciplinary Tumour Boards in Switzerland: Quo Vadis? Swiss Med Wkly 138(9–10):123–7
[263] Morimura K et al. (2004) Possible Distinct Molecular Carcinogenic Pathways for Bladder Cancer in Ukraine, before and after the Chernobyl Disaster. Oncology Reports 11(4):881–886
[264] Plattner H (2013) A Course in In-Memory Data Management: The Inner Mechanics of In-Memory Databases. Springer
[265] Singh J (2011) Understanding ETL and Data Warehousing: Issues, Challenges and Importance. Lambert Academic Publishing
[266] World Health Organization (2012) International Classification of Diseases. http://www.who.int/classifications/icd/en/. Accessed Sep 23, 2013

List of Abbreviations

AACR .... American Association for Cancer Research
AIS .... Active Information Store
API .... Application Programming Interface
ARIS .... Architecture of Integrated Information Systems
BI .... Business Intelligence
BioPAX .... Biological Pathway Exchange
BPMI .... Business Process Management Initiative
BPMN .... Business Process Model and Notation
CDSS .... Clinical Decision Support System
CEM .... Clinical Element Model
CGUL .... Custom Grouper User Language
CLOB .... Character Large Object
ConsensusPathDB . Consensus Pathway Database
CPU .... Central Processing Unit
CSV .... Character Separated Values
DGV .... Database of Genomic Variants
DNA .... Deoxyribonucleic Acid
DT .... Design Thinking
EC2 .... Amazon Elastic Compute Cloud
EHR .... Electronic Health Record
EMR .... Electronic Medical Record
EPC .... Event-driven Process Chain
ETL .... Extract, Transform, Load
FCFS .... First Come First Served
FCS .... Functional Class Scoring
FDA .... Food and Drug Administration
FMC .... Fundamental Modeling Concepts
FSB .... Front Side Bus
GDP .... Genome Data Processing
GP .... Gene Points
GTDS .... Gießener Tumor Documentation System

HG	Human Genome
HGNC	HUGO Gene Nomenclature Committee
HIG	High-performance In-memory Genome Platform
hiPathDB	Human-integrated Pathway Database
HIS	Hospital Information System
I/O	Input/Output
ICA	Independent Component Analysis
ICD	International Code of Disease
IMDB	In-memory Database
JSON	JavaScript Object Notation
KEGG	Kyoto Encyclopedia of Genes and Genomes
KGML	KEGG Markup Language
MIF	Molecular Interaction Format
MTHSPL	Metathesaurus Structured Product Labels
NCBI	National Center for Biotechnology Information
NER	Named Entity Recognition
NGS	Next-generation Sequencing
NLP	Natural Language Processing
NUMA	Non-uniform Memory Access
OMG	Object Management Group
OMIM	Online Mendelian Inheritance in Man
ORA	Over-representation Analysis
PAL	Predictive Analysis Library
PDF	Portable Document Format
PID	Pathway Interaction Database
PTBA	Pathway Topology Based Approach
QPI	Quick Path Interconnect
RNA	Ribonucleic Acid
SAM	Sequence Alignment/Map
SBML	Systems Biology Markup Language
SD	Standard Deviation
SHARP	Strategic Health IT Advanced Research Projects
SIF	Simple Interaction Format
SME	Small and Medium-sized Enterprise
SOM	Self-organized Maps
SPL	Structured Product Labeling
SQL	Structured Query Language
STF	Shortest Task First
UCSC	University of California, Santa Cruz
UDP	User Datagram Protocol
UI	User Interface
UML	Unified Modeling Language
UMLS	Unified Medical Language System
URI	Uniform Resource Identifier
VCF	Variant Call Format

WFMC	Workflow Management Coalition
WIPE	Weakly-structured Information Processing and Exploration
XML	eXtensible Markup Language
XPDL	XML Process Definition Language

Index

G

H

I

K

M

N

O

P

S

MIX
Papier aus verantwortungsvollen Quellen
Paper from responsible sources
FSC® C105338

If you have any concerns about our products,
you can contact us on
ProductSafety@springernature.com

In case Publisher is established outside the EU,
the EU authorized representative is:
Springer Nature Customer Service Center GmbH
Europaplatz 3, 69115 Heidelberg, Germany

Printed by Libri Plureos GmbH
in Hamburg, Germany